The West Point Military Library

General Editors of the Series

COLONEL THOMAS E. GRIESS
Professor, United States Military Academy

and

PROFESSOR JAY LUVAAS
Allegheny College

THE

SPIRIT

OF

MILITARY INSTITUTIONS,

BY

MARSHAL MARMONT,

DUKE OF RAGUSA.

TRANSLATED FROM THE LAST PARIS EDITION (1859), AND AUGMENTED BY BIOGRAPHI-
CAL, HISTORICAL, TOPOGRAPHICAL, AND MILITARY NOTES; WITH A NEW VERSION
OF GENERAL JOMINI'S CELEBRATED THIRTY-FIFTH CHAPTER, OF PART I,
OF TREATISE ON GRAND MILITARY OPERATIONS.

BY

FRANK SCHALLER,

COLONEL 22D REGIMENT MISSISSIPPI INFANTRY, CONFEDERATE ARMY.

GREENWOOD PRESS, PUBLISHERS
WESTPORT, CONNECTICUT

Library of Congress Cataloging in Publication Data

Marmont, Auguste Frédéric Louis Viesse de, duc de
    Raguse, 1774-1852.
    The spirit of military institutions.

    Translation of De l'esprit des institutions mili-
taires.
    Reprint of the 1864 ed. published by Evans and
Cogswell, Columbia, S. C.
    1. Military art and science. I. Schaller, Frank,
ed. II. Jomini, Henri, baron, 1779-1869.
III. Title.
U102.M313 1975   355   68-54799
ISBN 0-8371-5018-3

Entered according to Act of Congress, in the year 1864, by
EVANS & COGSWELL,
In the Clerk's Office of the District Court of the Confederate States for the District
of South Carolina.

Originally published in 1864 by Evans and Cogswell, Columbia, S.C.

Reprinted in 1974 by Greenwood Press,
a division of Williamhouse-Regency Inc.

Library of Congress Catalog Card Number 68-54799

ISBN 0-8371-5018-3

Printed in the United States of America

INDEX.

Dedication to President Jefferson Davis........................ 7
Letter of Colonel Schaller to President Davis................... 9
Reply of President Davis....................................... 11
From the Translator.. 13
List of Works used in the preparation of Notes................. 15
Dedication of Marshal Marmont's work to the Army of France.... 19
Some Introductory Advice to the Military Student............... 21

PART FIRST.

GENERAL THEORY OF THE MILITARY ART.

CHAPTER I.

Definitions... 27

CHAPTER II.

General principles.. 31

CHAPTER III.

Bases, Lines of Operation, and Strategy....................... 34

CHAPTER IV.

Tactics... 41

CHAPTER V.

Manœuvres... 44

PART SECOND.

ORGANIZATION, FORMATION, AND MAINTENANCE OF ARMIES.

CHAPTER I.

Organization and formation of troops.......................... 47
First Section: Infantry....................................... 47
Second Section: Cavalry....................................... 56

4　　　　　　　INDEX.

CHAPTER II.

Artillery.. 68

CHAPTER III.

Fortifications.. 98

CHAPTER IV.

Administration... 114
First Section: Subsistence... 115
Second Section: Hospitals.. 118

CHAPTER V.

Military justice and composition of courts......................... 122

PART THIRD.

DIVERS OPERATIONS OF WAR.

CHAPTER I.

Employment of the different arms...................................128

CHAPTER II.

Offensive and defensive wars......................................141

CHAPTER III.

Marches and encampments...147

CHAPTER IV.

Grand reconnoissances, and precautions they require...............156

CHAPTER V.

Detachments in presence of the enemy—their chances, and the
dangers which accompany them 158

CHAPTER VI.

Battles...162

CHAPTER VII.

Conduct of the general the day following the victory..............177

CHAPTER VIII.

Retreats 181

CHAPTER IX.

Night attacks and surprises..185

CHAPTER X.

Defence of fortresses......................................,........200

PART FOURTH.

PHILOSOPHY OF WAR.

CHAPTER I.

Morals of soldiers, and how to form them. Former armies, and
those of the present time ....＊. 20

CHAPTER II.

Military spirit, and difficulties of commanding.................. 223

CHAPTER III.

Picture of a general who answers to all the requirements of the
command... 232

CHAPTER IV.

Reputation of generals..................................... 260

Conclusion 262

APPENDIX.

Exposition of the general principles of the Art of War........... 26

THIS VOLUME IS INSCRIBED

TO

HIS EXCELLENCY

JEFFERSON DAVIS,

PRESIDENT OF THE CONFEDERATE STATES OF AMERICA,

AND

COMMANDER-IN-CHIEF OF THEIR ARMIES AND NAVIES,

WITH SENTIMENTS OF THE GREATEST

ESTEEM.

LETTER OF COLONEL SCHALLER TO HIS EXCELLENCY, PRESIDENT DAVIS.

COLUMBIA, SOUTH CAROLINA,
November 10, 1863.

His Excellency, President JEFFERSON DAVIS,
Richmond, Virginia.

MR. PRESIDENT: Every soldier of the Confederate States Army, of whatever grade, who is imbued with the spirit of advancement and emulation in his profession, has long since felt the want of the material which might serve him as a basis for both the study of, and research into, military matters.

Struggling as we are—alone, isolated, and without any exterior aid from any quarter of the world—presenting the greatest moral spectacle, and, in the exhibition of our unity and perseverance, the most graphic instance of a nation's devotion to the cause of liberty history has ever furnished—the soldiers of this Confederacy, amid their manifold privations, have not even had the gratification of comparing their illustrious deeds with those of other military nations, related to us by prominent military writers, and the study of whose works forms so important a part of a soldier's education.

The want of a work on the art of war, which could be easily comprehended by one even who had devoted but little time or thought upon military matters, and which yet would not be beneath the consideration of the educated soldier, being seriously felt, it was often brought to the notice of the writer.

During the period of my incapacity to serve in the field, arising from the effects of severe wounds, I have sought to prosecute, and I trust to have done so with some advantage, the study of military matters.

Consulting those of my professional friends to whose opinion consideration must be accorded, among several recent French and German works upon the military art at my disposal none struck me as worthier of being known to the great body of the Confederate Army than the work of Marshal Marmont, entitled: *"De l'Esprit des Institutions Militaires,"* Paris, 1859; and I have, in consequence, completed a translation thereof, with such military, historical, and topographical notes of my own, as appeared to be necessary for its complete understanding.

Concise, and of comparatively small compass for so extensive a subject as is presented by the consideration of the Art of War, it nevertheless forms a complete treatise upon it, and every important principle of that art "which elevates and preserves empires," and through the agency of which we will have to achieve our independence, is clearly and fully illustrated.

It is proper here to state that I have, in my notes, carefully abstained from criticising any of the operations of any of our generals, although reference has been made to illustrate, through the events of our own war, yet fully impressed upon the minds of our own soldiers, the principles of the art.

When once our independence is achieved, and the tableau of our successes and reverses is spread before us in all its glories and shadows, then, and only then, it appears to me, can a thorough criticism of our military operations be useful to the country and beneficial to the military student.

To you, Mr. President, whose genius has not only *organized* an army, but has achieved the still greater task of *maintaining* and *strengthening* it, amid circumstances the most unfavorable, until it has become, in numbers, spirit, devotion, and efficiency, more formidable than it has ever before been, even after a period of over thirty months of active warfare—to you, then, whose great achievements in this respect alone make you a master of the art of war, I have proposed, and herewith most respectfully beg permission, to dedicate my work.   *   *   *

With sentiments of the highest esteem,

I remain, your Excellency's

Most obedient servant,

FRANK SCHALLER,

Colonel 22d Regiment of Mississippi Infantry.

REPLY OF HIS EXCELLENCY, PRESIDENT DAVIS, TO COLONEL SCHALLER.

CONFEDERATE STATES OF AMERICA,
EXECUTIVE DEPARTMENT,
Richmond, Va., Nov. 25, 1863.

DEAR SIR: It gives me pleasure to acknowledge your letter of the 10th inst., and thank you for the personal kindness therein expressed.

The work of Marshal Marmont, Duc de Ragusa, *"De l'Esprit des Institutions Militaires,"* is an interesting one, and a good translation. of it will add something to our military literature.

In the 1st and 2d volumes of the United Service Journal, published in England in 1845, will be found a translation of this work, but evidently by one not skilled in the knowledge and use of the English language. Another translation, therefore, will be timely and useful.

The request you have been pleased to make in relation to the dedication, I grant, of course, with pleasure.

Very truly and respectfully yours,
JEFFERSON DAVIS.

COLONEL F. SCHALLER, etc., Columbia, S. C.

In offering to the officers and soldiers of the Confederate Army this little volume, I refrain from making any excuses. Many of them will be better able than myself to judge of its intrinsic merits.

It will be seen that, as an appendix, I have incorporated with it the lucid and justly celebrated chapter of General Jomini, giving, in a masterly manner, the exposition of the general principles of the art of war. I have sought to preserve its original force and integrity as much as my knowledge of the English language enabled me to do, and I believe it to be a faithful version. It is intended to be the forerunner of the entire work, which, as soon as the labors shall be finished, will appear in a form worthy of the illustrious author. Nothing but the great want of material, which is felt so much by our officers, has determined me to detach it, and to publish it in advance; its perusal will be attended with profit to soldiers of all ranks.

I can only express the hope that my companions-in-arms may find, in this little volume, some further encouragement steadfastly to advance in the glorious contest in which they are engaged, and that the arms of the Confederacy may soon be surmounted by the crown of Peace, as the finishing feature of their laurel and cypress-entwined glories.

For the defects of the work, as chargeable to myself, I would bespeak indulgence.

2

LIST OF WORKS USED IN THE PREPARATION OF NOTES.

1. Annales des Provinces Unies, par Basnage. A la Haye: 1726.

2. Traité des Grandes Opérations Militaires; par le Lieutenant-Général Jomini. Paris: 1818.

3. History of the War in France and Belgium in 1815; by W. Siborne. London: 1818.

4. Histoire Critique et Militaire des Guerres de la Révolution; par le Lieutenant-Général Jomini. Paris: 1820.

5. Atlas des plus mémorabiles Batailles, Combats, et Siéges ; par Fr. de Kausler, Major à l'État-Major-Général Wurtembergeois. Carlsrouhe et Fribourg: 1831.

6. History of the late Polish Revolution; by Major Joseph Hordynski, Polish Army. Boston: 1833.

7. Military Technical Dictionary; by Captain Burn, R. Artillery. Woolwich: 1842.

8. A View of the Formation, Discipline, and Economy of Armies; by Robert Jackson, M.D., Inspector-General of Army Hospitals. London : 1845.

9. Géographie, Physique, Historique, et Militaire; par La Vallée. Paris: 1850.

10. Military Maxims of Napoleon; translated from the French, by Lieutenant-General Sir G. C. D'Aguilar, C.B. London: 1852.

11. Military Encyclopædia; by J. H. Stocqueler. London: 1853.

12. On Fire-arms; by Colonel Chesney, R. Artillery. London: 1853.

13. On Cavalry; by Captain Nolan, 15th Hussars. London: 1854.

14. History of Europe; by Sir A. Alison. London: 1857.

15. Histoire de la Campagne de 1815; par Lieutenant-Colonel Charras. Bruxelles: 1858.

16. U. S. House of Representatives, 35th Congress, 2d Session. Ex. Doc. No. 2. Washington: 1858.

17. Der Italienische Krieg, 1859; von W. Rüstow. Zürich: 1859.

18. L'Italie Confédérée. Histoire Politique, Militaire, et Pittoresque de la Campagne de 1859; par A. de Césena. Paris: 1859.

19. U. S. Senate, 36th Congress, 1st Session. Ex. Doc. No. 2. Washington: 1860.

20. Hand-book Dictionary; by Lieutenant-Colonel Percy Scott, Commandant Isle of Wight Artillery. London: 1861.

21. Official Reports of Battles; published by order of Congress C. S. Richmond, Va.: 1862.

22. The Practice of War; by C. F. Pardigon.   Richmond, Va.: 1863. [Translation.]

23. Histoire et Tactique des trois Armes, et plus particulièrement de l'Artillerie de Campagne; par Ild. Favé, Capitaine d'Artillerie.   Paris: 1853.

THE SPIRIT

OF

MILITARY INSTITUTIONS.

2*

DEDICATION OF MARSHAL MARMONT'S WORK TO THE ARMY OF FRANCE.

To the Army I dedicate my work. It has been my cradle. In its ranks I have passed my life. I have constantly partaken of its hardships, and more than once I have shed my blood during the heroic times, the memory of which will never be lost.

Arrived at that age at which all the interests and consolations of a lifetime are to be found in the meditation upon the past, I address to the army this last souvenir.

Those soldiers, my companions in arms, possessed every military virtue. To bravery and the love of glory, so natural to Frenchmen, they joined great respect for discipline, and boundless confidence in their chief—the first elements of success.

While under my command, and when opposed to equal forces, they have never been beaten. Often victors, despite the inferiority of numbers, they but seldom gave way to immense superiority in strength and the fatality of circumstances; always remaining sufficiently formidable to make the enemy almost regret his victory.

The soldiers of the present time worthily follow the examples of their predecessors; and the courage, patience, and energy they unceasingly show during the long and painful African wars, prove that always, and everywhere, they will respond to the wants and the exigencies of the country.

The former have been the object of my most assiduous cares and my liveliest solicitude.

The latter, as long as I shall live, will have my most ardent sympathies.

THE MARSHAL, DUKE OF RAGUSA.

SOME INTRODUCTORY ADVICE TO THE MILITARY STUDENT.

Scarcity of military works—Special treatises do not contain the principles of the subject—Character of ancient writers—Polybius and Vegetius more curious than useful—Character of actual war—Difficulties of command in modern warfare—In what it consists—Alexander, Hannibal, Cæsar—Dreams of Knight Folard—Camp of Bayeux—Experience of Dumesnil-Durand—General Rogniat—Particular object of this book—Wealth of military literature—Standard works—Napoleon's writings—Archduke Charles' principles of strategy—Marshal Gouvion St. Cyr—Twenty campaigns, and fifty years of experience.

Notes.—1. Polybius. 2. Vegetius. 3. Gen. A. S. Johnston, Confederate Army. 4. Chevalier Folard. 5. Lieutenant-General Rogniat. 6. Thersites. 7. Memoirs of Montholon. 8. Archduke Charles of Austria. 9. Marshal Gouvion St. Cyr. 10. Ségur.

No modern work upon the art of war or military institutions contains a complete exposition of principles. Several special treatises upon the different arms have been published, but the principles of the subject at large have never been established in any. Superficial views, technical and minute details, without any sufficient indication of the aim and the means, can there only be found.

Ancient writers have searched into military questions even more profoundly; but what remains of their theories, since, through the discovery of gunpowder, military science has been so completely modified?

Polybius[1] and Vegetius[2] can still satisfy our curiosity; but we may no longer look in their writings for any useful and connected instruction.

Ancient and modern wars have no point of resemblance, their moral affinity perhaps excepted, or that sublime part of the art which consists in the knowledge of the human heart—a knowledge which, at all times, is so necessary in directing men, and which, in war, exercises a yet more prompt and decisive influence.

Everything has changed in the form and proportions of arms; their greater range keeps the combatants at greater distances; they cause greater terror, and contribute materially to prodigious results.

To this may be added the smaller number of combatants in former times.

But the art of commanding offers nowadays many more difficulties. With the ancients, who fought always with the hand, the army was formed in a compact manner; the small number of soldiers occupied but a very limited space; its front barely equalled that of one of our brigades. If a general could not himself see every one, he was at least within sight of all of his men. Operating upon so small a scale, the supreme chief could hasten everywhere; he was a combatant himself, giving the example, sword in hand. In our days, the general wages war by means of Will and Thought; his skill in wielding the sword is without importance; the mind embraces quite another range than is apparent to our perception; and a general, in short, is nowadays much less of a soldier, although he should now and then be one, but rather a moral being who, by force of his influence upon the understanding, appears to govern events like the mysterious powers of nature.[3]

Thus, actual war constitutes an entirely novel art, for which neither model nor precept could be found in the wars of the Greeks and Romans.

If the greatest captains of antiquity, Alexander, Cæsar, or Hannibal, could return to this earth, and come suddenly upon a field of battle, their genius would comprehend nothing; and they would have need of more than one campaign to understand completely the mechanism of the profession, as well as the consequences of our new arms and institutions.

These truths are so evident to him who has led in war, that it may well be asked how, in the times of Louis XIV, the reveries of Chevalier Folard[4] could have been seriously entertained; and later, the still more foolish vagaries of Ménil-Durand, who was on the point of establishing a special camp at Bayeux to obtain comparative practical knowledge of formations and manœuvres. It will be still more astonishing that a general of our epoch, Lieutenant-General Rogniat,[5] of the engineers, an officer of merit in his special arm, should have written a large book to revive and enlarge upon these vagaries; and if he did not figure among combatants, he could and might at least have seen some battles.

I propose to give a summary of the principles of organization, military institutions, and of active operations. I have endeavored to show that nothing should be left to hazard in these matters; that everything depends upon a generating principle, from which the necessary consequences flow.

A principle is discovered by the careful consideration of the end, and by seeking the best means to attain it.

A genius applies recognized principles; in this consists the whole art of war.

I have deemed it useful to expose them in the simplest manner, and to embrace, in a treatise on the rudiments of the military art, all the branches of that art, and at the same time the different services of armies, by stripping them of the technical charlatanism which too often accompanies them.

Yet to these principles the studies of military men are not limited; they should read attentively, in addition, the history of the campaigns of great generals—because the whole genius of these superior men is to be found in their application.

In this respect military literature is very rich; but a choice must be made. It is preferable to go back to the source; the works of those, therefore, who have commanded, should be sought; because little good is to be gained from campaigns described by subalterns, who, strangers to all the difficulties of command, and often to the very first notions of the profession, set themselves up for masters by censuring; they are modern Thersites,[6] severe in language but feeble of heart and arm, rather made to talk than to act. Their works are a tissue of errors and misrepresentations.

Above all documents deeply to be studied, the papers dictated by Napoleon, and published under the title of *"Memoirs of Montholon,"*[7] are recommended.

In every line the superior genius, the power of reasoning, and the authority of the great captain, are recognized. His decisions and explanations, though sometimes susceptible of controversy, offer the most valuable precepts; he who knows how to meditate upon and comprehend them will have the instinct of war.

An older work, which can not be studied too much, is the book published by the Archduke Charles of Austria, under the title of *"Principles of Strategy."*[8] In it the application of these principles is seen in his movements, when operating in 1796 against the Armies of the Rhine and Sambre-and-Meuse; it is an exemplification of all the rules of grand warfare.

The Memoirs of Marshal Gouvion Saint Cyr,[9] and The History of the Russian Campaign,[10] by Ségur, will likewise be read with advantage. From such sources healthy instruction, or the most correct opinions, can alone be drawn.*

I have occupied myself with the constitution of the different arms, and their best employment, for a long time; and I believe the principles which I am about to lay down to be true. I recommend them to the

---

* I visited, in 1826, during an entire day, the field of battle of the Moscowa, with several French and Russian officers who had participated in that struggle; upon the ground I read the three descriptions by Ségur, Chambray, and Boutourlin; in my opinion, the first gives the only correct account of the manner in which the events must have occurred.—*Note of Author.*

ardent, intelligent, and valorous youths who have replaced us; I have written for them.

The work I publish is the last contribution which, at the decline of my life, I am able to offer for the benefit of a science I have always cultivated with ardor, and a profession to which I am passionately attached.

In the moments of my leisure I found a great charm in the preparation of this resumé of my past studies and recollections. It is the fruit of my meditations, developed by long and frequent conversations with Napoleon, twenty campaigns, and more than half a century of experience.

NOTES.

1. **Polybius,** Greek historian; scholar of Philopœmen; friend and counsellor of Scipio Æmilianus; commander of the Achæan cavalry; defender of Greek independence. Among his works the military reader will find a treatise of the Roman Art of War, and another on Tactics. Died, of a fall from his horse, B.C. 121, 82 years old. His works have been translated into English by Hampton.

2. **Vegetius,** Roman military writer of great celebrity under the reign of the Emperor Valentinian, A.D. 386. His works may be found in 8 vols., published at Strasbourg, 1806.

3. **A general, in short, is nowadays much less of a soldier, although he should now and then be one.** This should be understood that a general, when occasion requires it, will not hesitate to perform the part of a soldier by joining in the combat. That he should be a soldier, in the sense of the term ordinarily applied, is presupposed. Our glorious General Albert Sidney Johnston, when leading Bowen's brigade to the charge to restore the wavering fortunes of the battle, at a moment which proved to be the turning-point of the struggle (Shiloh, April 6, 1862), forcibly illustrates this remark, and his act and heroic death will not only be one of the most magnificent passages of our history, but a sublime precept to every Confederate general. [*Vide* Chapter 3, Part IV.]

4. **Chevalier Charles de Folard,** born at Avignon in 1669; died there 1752; second lieutenant in the regiment of Berry, 1688; distinguished in campaign of 1701; thrice wounded in the Battle of Cassano, in 1705, gained by Louis, Duke of Vendôme, over the Imperialists, in the war of the Spanish succession. In 1714 he fought at Malta against the Turks; joined King Charles XII of Sweden, remaining with him to his death. His last campaign was with Marshal Berwick against Philip V of Spain, in 1719. In his Commentaries on Polybius he has given his views at length, especially his system of columns, for which he will chiefly be interesting to military men. His other principal works are: *Nouvelles Découvertes sur la Guerre; Traité de la Défense des Places, and Traité de la Guerre de Partisan.*

5. **Lieutenant-General of Engineers Rogniat** has published a very interesting work, entitled "*Considérations sur l'Art de la Guerre.*" In it he gives a complete exposition of the Roman art of war, and especially treats of the organization and tactical formation of the army of that celebrated military people,

recommending the legionary formation as being suited for our times. His work is written in a very philosophical manner, and quite interesting—in some respects, instructive.

6. Thersites, one of the Greeks in the army before Troy. Homer describes him as equally deformed in person and in mind. Such was his propensity to indulge in contumelious language, that he could not abstain from directing it against, not only the chiefs of the army, but even Agamemnon himself. He ultimately fell by the hand of Achilles, while he was ridiculing the sorrow of that hero for the slain Penthesilea.

7. The Memoirs of Montholon (Mémoires de Napoléon, dict. au Montholon et Gourgaud, 7 vols., London, 1823) are a perfect treasure to the military student, resplendent with the genius of its illustrious author. They were dictated to the two generals, Montholon and Gourgaud, who voluntarily accompanied the emperor to St. Helena, and are generally known under the name of the former.

Count Charles Tristam de Montholon, born at Paris, 1783. One of the noblest characters the period in which he lived has produced. The son of a soldier, and himself one from his fifteenth year, he served under Bonaparte on the 18th Brumaire (8th of November, 1799); subsequently aide-de-camp to Marshal Berthier; distinguished himself at Austerlitz, Wagram, Jena, and Friedland; specially attached to Napoleon. Commandant of the Department of the Loire, 1814; general in 1815; accompanies the emperor to St. Helena; edits his memoirs; accompanies the present emperor to Boulogne on the 6th of August, 1840; imprisoned in France.

General Gaspard Gourgaud, born at Versailles, 1783; educated at Polytechnic School; Teacher of Fortification at Chartres and Metz. 1801, enters 6th regiment artillery; in 1805, wounded at Austerlitz; distinguished in Battles of Jena, 1806; Poland, 1807; Siege of Saragossa, 1808; Battles of Abensberg, Eckmühl, Ratisbon, Ebersberg, Esslingen, and Wagram, 1809. Director of armory at Versailles; in Russia, 1812; at Lützen and Bautzen, 1813; saved the emperor's life at Brienne, 1814; distinguished at Nangis, Laon, and Rheims; with the Bourbons until their flight; again joins Napoleon; adjutant-general after the Battle of Ligny; one of the last at Waterloo, 1815; accompanies Napoleon to St. Helena; returns in 1818, to Europe; publishes an account of the Battle of Waterloo, for which arrested and banished; assists, on 15th of December, 1840, at the reinterment of Napoleon in the Church of the Invalides, where he placed the hat on the emperor's coffin.

8. Archduke Charles has given to the military student two works, from which the greatest benefit may be derived. They are: 1. *Grundsaetze der Strategie erlauetert durch die Darstellung des Feldzuges von* 1796, *in Deutschland.* (Principles of Strategy, illustrated by the Campaigns of 1796, in Germany.) Vienna: 1813. Five volumes, with a map of the theatre of war, and eleven plans. And, as a continuation of the same, 2. *Die Geschichte des Feldzuges von* 1799, *in Deutschland und in der Schweiz.* (History of the Campaign of 1799, in Germany and Switzerland.) Vienna: 1819. Two volumes, with an atlas in folio.

"His memoirs are models of lucid and authentic military history, worthy to be placed beside the *Commentaries* of Cæsar or the *Reveries* of Marshal Saxe. The principles of strategy on a great scale, to which the greatest successes or reverses in war are to be ascribed, never were more profoundly reflected on, or lucidly explained, than by this great commander. Like the Dictator, he discusses his own measures with an impartiality which is, literally speaking, *à toute épreuve.* To the merits of others, and, most of all, his opponents, he is ever alive, and yields a willing testimony; he is silent only on the praise due to his own great achievements."—*Alison.*

He was born in 1771, and died in 1847. In 1793, commands in Brabant the vanguard of Prince of Cobourg; Governor of the Netherlands. 1796, field-marshal of the German empire; beats Moreau; routs Jourdan at Amberg, Würtzburg, etc.; forces Jourdan and Moreau to retreat over the Rhine; takes possession of the fortress of Kehl, in winter of 1797; in 1799, defeats Jourdan in Suabia; opposes Massena in Switzerland; 1800, Governor-General of Bohemia; 1801, Minister of War; 1805, commands against Massena, and beats him at Caldiero; commander-in-chief of army of Austria; 1809, is beaten by Napoleon in Bavaria; shortly after beats Napoleon at Aspern; is beaten at Wagram; 1814, Governor of Mentz and Captain-General of Bohemia. Lives in retirement until his death.

9. The works which Marshal St. Cyr has written are: 1. *Guerres de 1792—1794.* 4 vols. Paris: 1829. 2. *Histoire Militaire,* 1799—1813. 4 vols. Paris: 1831. 3. *Guerre en Catalogne.* Paris: 1829.

He was born at Toul, a town in the Department of Meurthe, and on the railroad from Strasburg to Paris, in 1784. Enlisted as a private and marched to the Rhine, and rose from grade to grade. General of division in 1795; with Massena in Italy, 1798; appointed to succeed Massena. In 1799, sent to the Rhine; commands third corps under Moreau. 1800, tardy at Battle of Mœskirch; beats the Austrians at Biberach; saves Moreau at Memmingen; retires from the army of Moreau; fights in Italy. 1801, ambassador to Madrid; commands Neapolitan army till 1805. 1805, colonel-general of cuirassiers; distinguished in campaigns of Prussia and Poland. 1807, Governor of Warsaw; sent into Spain; superseded by Augereau; exiled two years. Greatly distinguishes himself at Polotsk in 1812, and is made a marshal. Commands at Dresden, and capitulates in 1813. Goes over to the Bourbons in 1814; made Peer of France; retires upon return of Napoleon from Elba; is made Minister of War by Louis XVIII; retires, and again Minister of War in 1818. Permanently retires in 1819. Died in March, 1830.

10. The work of Segur is entitled: *Campagne de* 1812. 2 volumes. Paris: 1825.

Although Count Ségur was not a military man, he has yet given an admirable military history of the disastrous campaign of 1812. His history is brilliant in the highest degree, and well worthy to be perused by the military student.

PART FIRST.

GENERAL THEORY OF THE MILITARY ART.

CHAPTER I.

DEFINITIONS.

Art of war—Genius of war—Necessity of knowledge of the human heart—Why a
man may be equal to ten, and ten men worth but one—Mind and character—
Military arts—Strategy, tactics, artillery, fortifications—Organization, manœuvres, administration—Dangers—Physiognomy of battles—Sacrifice of life to some
good purpose—Profession of arms.

Notes.—1. Lieutenant-General Jackson, Confederate Army.   2. Rogniat on
Bravery.

Before entering upon the subject I shall begin with some definitions:

The *Art of War* is the whole of the necessary knowledge to conduct
a mass of armed men, to organize, move, and lead it to battle, and to
give the greatest valor to the elements composing it, while watching
over their preservation.

The *Genius of War* consists in the talent to apply these elements at
the right time and in the most suitable manner—to devise the best combinations with surety and promptitude, in the midst of dangers and
critical situations.

The genius of war is incomplete, if to the faculty of these combinations, and which I shall call technical, a general does not join the
knowledge of the human heart—if he has not the instinct to divine what
passes in the souls of his soldiers and those of the enemy.   These variable inspirations form the moral aspect of war; they are that mysterious
action, which, imparting sudden power to an army, makes one man
worth ten, and ten men worth but one.[1]

Two other faculties are equally necessary—authority and decision.
They are both gifts of nature.

A great general, then, must possess vast intelligence—but, still more,
character.   His character directs the execution; it alone, both in modern times and in antiquity, has made generals of the first order.

Military Arts consist in the knowledge of scientific or mechanical
transactions, which regulate the details of execution and the employment of the means.

Thus, strategy, tactics, artillery, fortifications, organization, and the administration of armies, are military arts with which a general should be familiar. Every art has its theory; but the talent of employing it with advantage demands frequent application and the spirit of observation.

Of all human events, those of war most especially claim, without contradiction, the assistance of the auxiliary called experience. We must try to accustom ourselves to the dangers of battle presented in such various forms. Man, born brave, will at first expose himself fearlessly to perils, sometimes with pleasure even; but time alone will teach him the faculty of appreciating the most useful manner in which to sacrifice his life.[2]

Finally, *the Profession of Arms* is life consecrated to military labors; and this definition applies particularly to those who execute.

NOTES.

1. **Lieutenant-General Jackson, Confederate Army.**—Our own struggle against the devastators of our country and the destroyers of the great undamental principle of self-government has produced one character who possessed the genius of war in a remarkable manner—Lieutenant-General Jackson. He certainly knew how to throw his troops upon the enemy at the right time and in the most suitable manner. If we can not as yet determine whether the original *conception* of his various able movements must be ascribed to him—a question which history alone must establish—the great merit is due him now, and will hereafter, in all times, be sufficient to illustrate his name—that he was by far the greatest *executor* this war has, as yet, brought to light. Not alone did he carry out all of his movements with the greatest decision and vigor, but his self-command in the midst of critical dangers, the smoke of battle, and the uncertainties of war, was something astonishing and assuring to all around him. His adjutant-general, Doctor Dabney, a Presbyterian divine, tells us that he found himself in a most critical situation at Coal Harbor on Friday morning, the 27th June, 1862; when moving forward, he did not know at what time the enemy might intercept his march, and thus frustrate the wisest combinations of the general-in-chief. Only a Jackson could have remained as steady and full of confidence under like circumstances.

But, however high we must place him as an able strategist and master of technical combinations, his peculiar gift was the knowledge of the human heart It is believed that he was one of the strictest disciplinarians in the service; his troops have made as hard marches and suffered as great privations as any, and yet did we ever hear that any of them did complain or murmur? It is only a genius who can bring about such results; those who fail to inspire their troops with a due degree of love and veneration are but ordinary men.

He gained the admiration of even our enemy. A fact which demonstrates that, even amid nations the most debased and corrupted, a spark of Godlike virtue will sometimes call forth an unwilling acknowledgment.

The following is the tribute of Governor Letcher, of Virginia, to his memory in his annual message to the legislature:

" Since the commencement of this war Virginia has been called upon to mourn over the loss of many of her gallant sons ; but of all her jewels, the most brilliant was the illustrious Lieutenant-General Thomas J. Jackson—a graduate of West Point—highly distinguished in the Mexican war—and, at the opening of the present war, a quiet, unpretending professor in our State Military Institute. He was called from the professor's chair to the field ; and his sagacity, his energy, and the unparalleled success which crowned his efforts, won for him a reputation that made him the pride of his own state—endeared him to the people of the Confederacy— attracted to him the attention of the nations of the earth, and compelled the respect and admiration even of those heartless enemies who have drenched our land in blood, and brought wailing and lamentation to the firesides of thousands of their own misguided people. For decision of character—for stern and unbending resolution—for pure and elevated patriotism—for sound and inflexible integrity, and for prompt and energetic action, he was surpassed by no man of his times. The record of his achievements will constitute some of the brightest pages in the history of the war, when that history shall have been written. His reputation as a military leader of the highest ability and merit has been fairly and firmly established in the judgment of the army and the country, and his name will be honored and his fame will be cherished

   ' While the earth bears a plant, or the sea rolls a wave.'

" General Jackson was not only a great man, but he was, emphatically, a good man. He was pure and upright, earnest and honest, conscientious and' true in his intercourse with the world. In all the relations of life—as a son, a husband, a father, and a citizen—he was faithful and reliable. As a member of the Presbyterian Church his ' walk and conversation' attested the sincerity of his profession. The death of such a man, and at such a time, could not fail to produce the most profound sensation throughout the Confederacy. He had won the confidence of the people of all classes. Their affections were entwined around him—their hopes centred in him—and they looked to him as one of the great instruments provided by an All-wise Providence for their deliverance, and for the establishment of their independence upon a new basis. His death was regarded as a national calamity, and it was succeeded by manifestations of the most heartfelt grief and the most sincere sorrow.

" He has passed from life, but his example is still left to encourage and stimulate us to greater exertions in the noble cause in which we are engaged. That example can not fail to exert a most powerful influence in awaking our dormant energies ; in rousing us up to greater efforts ; in inspiring us with greater zeal, and in animating us with a nobler spirit and a more determined courage. His whole soul was in the cause, and he performed his duty cheerfully and with the most scrupulous fidelity. The redemption of the people from the yoke of Yankee tyranny was the object nearest his heart, and to its accomplishment he directed his efforts· The cheeks of the deserters, and stragglers, and laggards should burn with shame when contemplating his devotion and his self-sacrificing spirit. They should appreciate such an example, and should resolve at once to emulate it, and should return to the path of duty with a fixed purpose to relieve their land from the tread of the invader, or, as he did, sacrifice their lives in the effort. If such shall be the result, he will not have died in vain. The sacrifice, great as it was, will impress upon the country an invaluable lesson for the instruction of the present and future generations."

   3*

And the great Thunderer, the English *Times*, thus announces his death, May 26th, 1863:

"The Confederate laurels won on the field of Chancellorsville must be twined with the cypress. Probably no disaster of the war will have carried such grief to Southern hearts as the death of General Jackson, who has succumbed to the wounds received in the great battle of the 3d of May. Even on this side of the ocean the gallant soldier's fate will everywhere be heard with pity and sympathy. Not only as a brave man fighting for his country's independence, but as one of the most consummate generals that his country has produced, 'Stonewall' Jackson will carry with him to his early grave the regrets of all who can admire greatness and genius. From the earliest days of the war he has been conspicuous for the most remarkable military qualities. That mixture of daring and judgment, which is the mark of 'heaven-born' generals, distinguished him beyond any man of his time. Although the young Confederacy has been illustrated by a number of eminent soldiers, yet the applause and devotion of his countrymen, confirmed by the judgment of European nations, have given the first place to General Jackson. The military feats he accomplished moved the minds of the people with an astonishment which is only given to the highest genius to produce. The blows he struck at the enemy were as terrible and decisive as those of Bonaparte himself. The march by which he surprised the army of Pope last year would be enough, in itself, to give him a high place in military history. But, perhaps, the crowning glory of his life was the great battle in which he fell. When the Federal commander, by crossing the river twelve miles above his camp, and pressing on, as he thought, to the rear of the Confederates, had placed them between two bodies of his army, he was so confident of success as to boast the enemy was the property of the Army of the Potomac. It was reserved to Jackson, by a swift and secret march, to fall upon his right wing, crush it, and, by an attack unsurpassed in fierceness and pertinacity, to drive his very superior forces back into a position from which he could not extricate himself except by flight across the river. In the battle of Sunday Jackson received two wounds, one in the left arm, the other in the right hand. Amputation of the arm was necessary, and the Southern hero sank under the effect of it, supported to the last by his simple and noble character, and strong religious faith."

2. Rogniat on Bravery.—Quite in opposition to Marshal Marmont, General Rogniat, in his twelfth chapter of his highly philosophical work, tells us that man, far from being born brave, is, on the contrary, a coward from the very first, and that he can only acquire the military virtue of bravery by experience; in short, that it is an artificial, not an innate, quality. He furthermore says: "Nature has given us instinctively the desire to preserve life, in the shape of a feeling of fear which evades everything that might become dangerous to it; and that courage consists in surmounting and conquering this sentiment."

Passions, he maintains, make us blind to dangers, even in the midst of the most evident perils; and they force us to brave all, in spite of the counsels of prudence. Why, then, are young men braver than the old, when the former have much more to lose than those whose course in life is nearly completed? Because passion influences the first more powerfully than the latter, in whose bosoms the fires of youth have given way to the icicles of old age.

Ignorance dares a danger for a moment. But experience soon enlightens, and fear succeeds security. Thus he saw young French recruits follow, at Lützen, their officers in the midst of perils unshrinkingly, and with less hesitation than at Bautzen, only a few days later. Experience had taught them the danger.

CHAPTER II.

GENERAL PRINCIPLES.

Very few—Infinite variety of circumstances in which an army may be found—
Combinations resulting therefrom — Hazard to be considered — Disproportion
between the genius of Napoleon and that of his adversaries —With equal
chances, victory will always be with him who knows how to make himself
superior in strength at a given moment — Inferiority in numbers compensated
by the qualities of troops — Barbarians always yield to disciplined troops less
numerous — After a first success, the spirit of troops replaces the weight of
arms — Greeks and Persians — Marathon and Platæa — Conquests of Alexander
— Romans against Germans and Gauls — French against Turks — Advantages of
taking the initiative — Campaign of Italy in 1796 and 1797 — Axiom.

NOTES.—1. Battle of Marathon. 2. Battle of Platæa. 3. General Wurmser.

There are but few general principles for the conduct of armies, but
their application calls forth a host of combinations which can neither
be foreseen nor classified as rules.

The conditions in which an army may find itself vary infinitely;
the principal points of importance are the mass of the elements composing it, the relative state of the two armies, the nature of the theatre
of war and of the adjoining countries, the part to be assumed, either
defensive or offensive, and the reputation and character of the general
against whom the field is taken, etc.

An immense field is open to combinations by divers circumstances—
so much so, that the vastest mind would not be able to embrace them
all. Hence, the greatest generals have committed faults; the best
generals are those who have been guilty of the least number. The
more elements are admitted in calculating these circumstances, the
better will events be controlled; generals should foresee the probable
as much as the possible; even accidental risks ought to be guarded
against. Thus, when the period of reverses is at hand, great catastrophes can be prevented.

This foresight was one of Napoleon's highest faculties during the
period of his fortune; his adversaries being almost destitute of it,
results were obtained by him at the time which astonished the world.

I shall establish some principles by which a general's conduct
should always be regulated. The aim I may indicate, but the means
of accomplishing the same must always be subordinate to circumstances.

Whenever two armies, nearly of equal strength and in the same
moral condition, engage each other, their chances of success are alike.
To render them more favorable, movements are planned to deceive the

enemy, and through fear he is led to divide his forces. Then the
general who is most skilful will promptly reassemble his forces to
overthrow his adversary's; any temporary superiority thus acquired
will greatly facilitate his victory.

Numerical superiority, at the moment of battle, is of extreme importance. Undoubtedly the quality of troops is to be considered
more than the number; but, in the present state of European armies,
numbers and the concentration of means contribute powerfully to
success. It is different when barbarians are to be met, who, devoid of
organization and without discipline, form no compact body; they
operate without unity and harmony; they are always inferior, at any
time, when opposed to the weakest but well-united body. Two repeated attacks without success, and often a single charge, will scatter the
less brave; the others follow the contagion of the example, and soon
all will be dispersed. Then the spirit of the troops replaces the arms.
The wars of the Greeks against the Persians, the Battles of Marathon[1]
and Platæa,[2] the conquests of Alexander, the triumphs of small Roman
armies against the Turks despite the disproportion of numbers, can all
be cited and explained in this connection.

When aiming to disperse the forces of the enemy, those points
necessary for his security should be particularly harassed; and whenever he gives way to appearances he must be attacked upon a weak
point with superior forces. This is precisely what, in fencing, is termed
a feint, when, sword in hand, we are engaged in single combat. Two
or three partial advantages prepare the way for more considerable
successes, which, in the end, will decide the fate of the campaign.

It will, therefore, be seen how important it is for a general to take
the initiative in his movements; he will then control the mind of his
adversary, and a first chance oftentimes gives an ascendency never
again lost. The favorable moment must be spied. A too great disproportion in forces and the different auxiliaries may be an insurmountable obstacle; in that case we must wait until the confidence of
the enemy leads him to commit some fault. Diligently improving the
occasion, a skilful general can obtain advantages which will permit
him to change *rôles*, and to pass from a defensive to an offensive attitude.

A notable instance of this kind took place in 1796, during the
immortal Italian campaign. The French army, having reached the
frontiers of the Tyrol in a defensive attitude, found itself greatly
inferior to the Austrian army, reinforced and led by Wurmser[3] in
person. When the hostile general attacked, his forces were divided;
the French general united his, and soon a first success permitted him,
in his turn, to assume the offensive. Then began a series of victories,

in which the French army maintained its ascendency in numbers upon almost every field of battle.

To resume, in a single word, this part of the art of war, which relates to the general movements of an army, it is to be observed that, in every instance, it rests upon a calculation as to the time required, the distance, and celerity of movement.

NOTES.

1. The Battle of Marathon was fought by the Persians and Athenians, near the seacoast of Greece, not far from Athens, B.C. 490. The Persians, as is most generally supposed, had an army of one hundred thousand infantry and ten thousand cavalry; the Greeks only eleven thousand men. Rather than endure the uncertainties and horrors of a siege, the Greek general, Miltiades, resolved to meet the enemy and attack him in the plains of Marathon. Taking a position at the foot of a mountain to guard his rear, he felled trees on both extremities of his wings to make the enemy's cavalry useless. At the expense of his centre, the wings had been considerably strengthened, and after the first onset of the Greeks, which was bravely received by the Persians under Datis, the latter took advantage of the apparent mistake of Miltiades, and endeavored to break his centre. But the Greeks, having foreseen this manœuvre, themselves gave way in the centre, and wheeling both of their wings, attacked the enemy simultaneously in front and rear, routing him completely. The loss of the Persians was six thousand four hundred; that of the Greeks not two hundred.

2. The Battle of Platæa was fought between the Persians, under Mardonius, some three hundred and fifty thousand strong, and the Greeks, about seventy thousand, under the Spartan, Pausanias, near the city from which it derives its name, in the province of Bœotia, in Hellas. We have no intelligible account of the tactics employed in this battle, which freed Greece for ever from Persian aggression; and it appears that the battle was rather gained by a judicious use of the ground than by any extraordinary bravery or skill of the Greeks. The latter, rent by internal dissensions in their army, arising from jealousies against the commander-in-chief, were, besides, disheartened by the want of water, and the continual harassing of the Persian troopers in an exposed country. Pausanias, therefore, resolved to retreat toward the city of Platæa. It being construed by the Persian general into a flight, he eagerly pressed forward, was led into an unfavorable position, and being at first charged by a gallant band of Spartans, the remainder of the Greek army were reassured, and the Persians not only were thoroughly beaten, but so massacred by the infuriated conquerors that not three thousand of them are said to have remained— Mardonius himself gallantly falling, while striving to restore the equilibrium of his troops. This battle teaches four lessons: 1. That everything in war depends upon a resolute front, and that there is no cause for despair while yet an army remains. 2. How eager an enemy is to fall into a trap, when he once imagines that the opposing army is demoralized. 3. The importance of the selection of the ground, and a thorough knowledge of the topography of the country in which we operate, and that a retreat or abandonment of ground is sometimes the prelude to victory. 4. That only with a total rout of the invader, such as the Persians experienced, may we hope to put an end to this war.

3. General Wurmser.—It was this general's misfortune to be placed agains a young giant, Napoleon, when he was already seventy-two years of age. Raised under Frederick the Great's tuition, though on the Austrian side, he could not fail to possess considerable talent. His operations in Italy, and his firmness in adversity, sufficiently show that he was not a general of ordinary merit. Born 1724; serves during the seven years' war; beats the Prussians at Habelschwerd, 1779; crosses the Rhine, March 31, 1793; compelled to recross, close of 1793; captures Manheim, November 22, 1795; commands Army of Italy, 1796; throws himself into Mantua, September 30, 1796; capitulates, 1797; soon after dies of exhaustion at Vienna.

CHAPTER III.

BASES, LINES OF OPERATION, AND STRATEGY.

Base of operation—Line of operation—Strategic points—Strategic lines—Chessboard of Napoleon—Consequences of violation of principles at Marengo—Moreau's strategy—His tactics—Hohenlinden—Character of that battle—To whom belongs the victory—General Richepanse—Importance of mobility in an army—Usage of Russians—Independence of reserves—Necessities relative to lines of operation—Liberty of communications—Base of operation—Its extent influencing line of operation—Fundamental axiom of Napoleon—Great success due its observance in 1805, 1806, 1809—Parallelism of bases of operations of opposed armies—March of 1812—Absence of base of operation—Admiral Tchitchakoff and General Kaptzievitch—Examples of useful changes of line and base of operation—1797, Austrian army after reduction of Mantua—1814, Marshal, Duke of Dalmatia, at Toulouse—Two principles.

Notes—1. Moreau. 2. March of French army. 3. Campaign of 1806. 4. Fortress of Mayence. 5. Bâle. 6. Beresina. 7. Generals A. S. Johnston, J. E. Johnston, and B. Bragg. 8. Mantua. 9. Soult. 10. Adour. 11. Toulouse. 12. Bordeaux.

The base of operation of an army is the country the forces protect, which furnishes their necessaries, which every day sends them the articles of all kinds they consume in men, horses, provisions, and ammunition, and which receives their sick and wounded, etc.

The line of operation is determined by the general direction of the march, which indicates the object of operation or the point to be attained.

General movements, executed beyond the sight of the enemy and before the battle, are called *strategy*.

Strategic points are those which it is important to occupy, be it to menace the communications of the enemy, or to cover our own. They should be chosen so as to facilitate the combinations of movements of the different columns of an army. Generally, a spot where many roads cross is a strategic point; in a mountain country, the spot where several valleys meet is a strategic point.

Strategic lines are those which unite divers strategic points, and which are used in the movements executed between the latter; they should be as short as possible.

The judicious choice of strategic points and lines is the safety of armies in reverses, and the cause of the greatest results in successes.

Napoleon particularly possessed the genius of strategy; no general has ever surpassed him in this respect; none knew better, beforehand, how to find the point where it was necessary to strike.

A large army is composed of several columns; they are necessarily separated in order to subsist and to move with facility. The most distant parts should be able to arrive in time for the battle, either to take part in the conflict, or to serve merely as a reserve. The object of strategy is to regulate the march for the promptest reunion upon the same point, be it at the centre or upon one of the wings. A march thus regulated is what Napoleon was in the habit of calling his chess-board.

All of his first campaigns have had this character, except at Marengo, where, deviating from this principle, he was about to succumb; upon the day of action he is always seen to reassemble, upon the field of battle, all the forces he could reasonably dispose of.

Moreau,[1] on the contrary, whose talents have been so highly praised, understood nothing of strategy. His skill consisted in the application of tactics. Personally very brave, in the presence of the enemy he handled troops well when they occupied a field which his view could embrace; but his principal battles he only fought with a part of his forces.

At Hohenlinden, where the success was so overwhelming, Moreau ought to have been defeated, and he probably would have been, if the Austrians had not manœuvred with a carelessness unexampled. The French army was composed of twelve divisions; the three of the right were commanded by General Lecourbe, and the three of the left, General Sainte-Suzanne commanding, did not assist in the battle. The Austrian army was united, but marched unconnectedly; the centre column, which encountered no obstacle, and followed the main road, with almost all the artillery, was open to an isolated attack, and not even formed; it was liable, at any moment, to be assailed in flank. This remarkable fortune was not due to the dispositions of General Moreau. General Richepanse, a man of mind and courage, finding himself with his division surrounded by the Austrians, who began to organize, faced about, and seized one hundred pieces of cannon, marching by column upon the high road.

The reunion of an army at the moment of the conflict being the object, and the rapidity of marches the means to obtain it, the divisions, which are the units, should be able to combine promptly, and, to that effect, be susceptible of very rapid movements. An army will have, under all circumstances, a slow march, but the rapidity should be imparted to the elements which compose it. Therefore the divisions should not be too much burdened with artillery and army stores. I do not approve the custom of the Russians, who encumber them with ordnance. The great reserves of material and army stores of every kind should have an independent march, able to defend themselves, and, when required, be escorted by special troops. It is the duty of the general-in-chief to keep them always within reach of the point where they can be most useful, according to their destination.

There is another object which should call forth the greatest solicitude of a general; he should have his line of operation covered perfectly, whilst menacing that of the enemy. Free communications are necessary for the maintenance of an army; once lost, the moral condition is compromised. Confidence—this powerful element of the mind, which nothing in the world can supply in troops—does not always resist such a trial.

Hence the necessity of a large base of operation. If a fortress and several fortified points are situated upon this base, or if a river forms part of it, great advantages may result. The more this base is extended, the better is the line of operation covered. This was one of Napoleon's fundamental axioms; he never violated it with impunity. In his splendid campaigns of 1805, 1806, and 1809, he has given great examples, and profited, with skill, by favorable circumstances resulting from the position of our frontier.

Two armies, which have parallel bases of operation equal in extent, are in like condition; and one of them, in turning the other, is also necessarily turned. But it is not the same when two bases of operation have different lengths, or are inclined toward each other.

In 1805, the French army, after the fine march from the shores of La Mancha to Germany,[2] was directed upon the flank and rear of the Austrian army which had invaded Bavaria. One battle lost upon the Danube would have thrown it back upon the Rhine; one battle gained overthrew the hostile arms.

In 1806[3] the French army, at the beginning of the campaign, found itself upon the flank of the Prussian army; its communications with France from Mayence[4] to Bâle[5] were not the less free; they were, besides, so well assured that a reverse could not have had any serious consequences, and one single victory brought about the results so well known.

In 1812, when Napoleon withdrew, without limitation, from his point of departure (for it is necessary to remark that the dimension of a base of operation, to satisfy the requirements, is not absolute, but relative to the length of the line of operation), his base disappeared. Established from the first upon the position of divers army corps, the base would have been sufficient, if the army had remained nearer the frontier. But these corps were abandoned to themselves upon the march, a prey to the chances of war; and meeting bodies of the enemy of at least equal strength, the sequel was, that the army lost all its communications. Arrived upon the banks of the Beresina,[6] Napoleon should have been thoroughly beaten, and the remnants of his army ought to have been annihilated, since there was no miracle required, and Admiral Tchit-chakoff and General Kaptzievitch could have attributed to themselves no credit whatever in that case.

But there are circumstances where it is useful and salutary to change, in the midst of a campaign, the direction of the line of operation, and to choose another base; and although the most natural idea and the most habitual usage would be to take a position in advance of the country to be defended, it yet sometimes happens that its security is in the most efficacious manner assured by taking a line of operation which appears to abandon it to the enemy.[7]

When, in 1797, after the reduction of Mantua,[8] the French army marched upon Vienna, the Austrian army, which found itself in an inferior condition to give battle, retired in the direction of the capital. If, instead of operating thus, it had taken position in the Tyrol, the natural obstacles presented in that country would have established a kind of equilibrium between the respective forces; the troops, consisting of new levies from Hungary and Croatia, which could not serve with utility on the day of battle, would have been sufficient to cover the frontier of Friuli, to hold in check a French corps, and paralyze its action, despite the excellence of its troops (because the French army was composed of troops of like nature). Furthermore, the Austrian army, in taking this line of operation, would have met efficient reinforcements, which it could have only received by the borders of the Rhine. Lastly, if the war had brought the belligerent armies into Suabia and Bavaria, all the Austrian forces, reunited upon the centre of operations, would have been able to manœuvre under the most advantageous circumstances. The Austrian army committed, then, a great mistake in adopting this line of operation.

Here is another example : In 1814 the marshal, Duke of Dalmatia,[9] after having operated upon the Adour,[10] was obliged to leave the basin of that river, and directed his line of operation upon Toulouse.[11] In this he acted wisely, because he thus removed the English army from the centre of France more surely than by retiring upon Bordeaux,[12] where it would have followed him; a small body of troops, sustained by the national guards, placed in rear of the heaths and covering Bordeaux, would have defended that city, provided the spirit of the epoch and internal political complications had not rendered these wise dispositions useless.

Reviewing the matter, strategy has a twofold aim:

1. To reunite all the troops, or the greatest number possible, upon the point of conflict, if the enemy has there but a portion of his; in other terms, to manage to secure a numerical superiority for the day of battle.

2. To cover and assure one's own communications, while all the time menacing those of the enemy.

NOTES.

1. Moreau, born at Morlaix, Bretagne, 1763; brigadier-general, 1793; general of division, 1794; commander of Army of the Rhine; retreats into France; commands Army of Italy, 1798; beaten by Suwarrow; commands, 1799, Army of Danube and Rhine; passes the Danube and Rhine; gives battles at Mœskirch, Engen, Memmingen, Biberach, Hochstaedt, Nördlingen, etc.; beats the Austrians at Hohenlinden, December 3, 1800; accused and tried for conspiracy; exiled; goes to America; settles at Morrisville, on Delaware river, 1805; returns to join the Allies; is mortally wounded before Dresden; dies, September 1, 1813; buried at St. Petersburg.

2. March of the French army from the shores of La Mancha to Germany.—An army of one hundred and eighty thousand marched, in from seventeen to twenty-two days, from Cherbourg, on the British channel, to the Rhine, to concentrate near Ulm, in Germany. The distance from Cherbourg to the Rhine is some five hundred miles. This was the splendid army which fought the campaign of Austerlitz, and was composed of ten corps, commanded by Bernadotte, Marmont, Davoust, Soult, Lannes, Ney, Augereau, Murat (cavalry), guards (Mortier and Bessières), and Wrede (Bavarians).

3. The Campaign of 1806 ended with the victories of Jena and Auerstaedt over the Prussians. The line from Mayence, in Germany, to Bâle, in Switzerland, is of exceeding importance.

4. Mayence, or Mainz (the ancient Maguntiacum metropolis of Germania Prima), is on the left bank, in the basin of the Middle Rhine, a very strong place in form of a semicircle, of which the Rhine is the diameter, covered by three fortified fronts to the north, the west, and the south, which include fourteen bastions and a double rampart, defended by six forts—the most considerable of which is Hauptstein, commanding all the environs; the isles of the Rhine are fortified, as well as the suburbs of Cassel, on the right bank. Mainz was formerly the capital of an ecclesiastical electorate; it was taken by the French in 1792, retaken by the King of Prussia in 1793, ceded to France by the treaty of Campo Formio, and remained in the possession of the latter till 1814, when it was given up to the Grand-Duke of Hesse-Darmstadt, and declared a fortress of the Germanic Confederation.

5. Bale, a city in the basin of the Upper Rhine, built on both sides of the river, capital of a Swiss canton, centre of the routes from Southern Germany into France, and having now a railroad to Dijon, another to Strasbourg, both in France, and a third along the right bank of the Rhine, through Freybourg, Kehl, Carlsruhe, and Manheim, to Frankfort-on-the-Main, and on to Cassel. Bâle is the true gate into France, and which is kept closed only by the neutrality of Switzerland. Passage of the Allies here in 1814 and 1815.

6. The Beresina, an affluent on the right bank of the Dneiper, rises in the marches of Dokchitsky, traverses a country of forests and bogs, where there is no advancing but upon raised causeways; flows past Studzianka, celebrated for the disastrous passage of the French army on the 26th of November, 1812, and for the double battle which they fought on that occasion with the Russians, on both banks of the river. Borisow, a town situated on the road from Wilna, by Minsk, to Smolensk, and the possession of which by the Russians compelled the French to seek a passage at Studzianka. After leaving this place the Beresina receives an affluent, which passes by Minsk, a town situated in the midst of extensive forests, on the road from Warsaw to Smolensk, and the possession of which by the Russians, in 1812, was one of the causes of the disaster of the Beresina. Had Napoleon directed his retreat upon Vitebsk, as he at first intended, and which he certainly would have done notwithstanding the reported bad state of the roads, had he known of Tchitchakoff's proximity to the Beresina, while he himself was at Orcha, the passage of the Beresina would have never been attempted. The passage of this stream cost the French twelve thousand killed and drowned, and sixteen thousand prisoners; and if Kutusoff had been more expeditious, and Admiral Tchitchakoff more prudent it is probable that neither Napoleon himself, nor a single man of his army, would have escaped.

7. Generals A. S. Johnston, J. E. Johnston, and B. Bragg.—This remark is forcibly illustrated by three events in our own war: 1. The splendid retrograde movement of General A. S. Johnston from the heart of Kentucky to the confines of the State of Mississippi, in February and March, 1862, which culminated in the Battle of *Shiloh*. 2. The withdrawal of General J. E. Johnston's army from the lines of Manassas to the lines of Richmond, which culminated in the Battle of *Seven Pines*, May 31, 1862, so glorious to the Confederate arms. 3. The abandonment of Tennessee by General Bragg, which resulted in the Battle of *Chickamauga*.

8. Mantua, situated among three small marshy and unhealthy lakes on the Mincio. It is the ancient capital of a sovereign duchy, and is now one of the strongest places in Europe. Its position on the left bank of the Po invests it with a degree of importance equal to that which attaches to Alexandria, on the right bank. It communicates with the mainland by five causeways, which are defended by five fortresses: the first of these, called *La Favorita*—battle here of 1797, gained by Napoleon over the Austrians—leads to Verona; the second, named *Santa Georgia*—battle in 1797, gained by Napoleon over the Austrians—leads to Legnano; the third, called *Del Pietoli,* skirts the Lower Mincio, and leads to Governolo; the fourth, called the *Cerese,* leads to Guastalla; and the fifth, called *De la Pradella,* leads to Cremona. From the lakes of Mantua a canal is drawn off to the south-west, which joins the Po, and forms, with the two streams, a triangular island of admirable fertility, which is called the Seraglio. This island is the great resource of the garrison of Mantua, which draws its provisions from it, and, by means of it, remains mistress of the course of the Po. The most celebrated of the many sieges which Mantua has experienced was that of 1796, in which Napoleon was obliged to destroy successively three Austrian armies before he could render himself master of the place. In April, 1848, Charles Albert, King of Sardinia, blockaded it, but was obliged to raise the siege; invested it again on the 13th of July, 1848, but was compelled, on the 26th of July, by the able movements of Radetzky, to abandon it. Is now incorporated with the Kingdom of Italy.

9. Marshal Soult fought, on the 10th of April, 1814, the Battle of Toulouse against Wellington, the last which was fought amid the tottering wreck of the Napoleonic dynasty. His admirable firmness and fidelity amid the universal gloom is above all praise. Military writers have censured him for taking the position of Toulouse, instead of that at Bordeaux, forgetting the political disturbances in the latter city at the time, and the necessity of surrender had he been unsuccessful in its defence.

10. The Adour, a river of France, rises in Mount Tourmatel, in the Pyrenees; falls in the Bay of Biscay, north of Bayonne.

11. Toulouse, head-quarters of the tenth military division; not now fortified; surrounded by an old, very thick wall, with flanking towers, only accessible from the south; situated upon the right bank of the Garonne. It has upon the left bank its suburb of St. Cyprien, built upon a creek, surrounded by a strong brick wall, which constitutes a good *tête de pont*. It is the point of convergence of the roads from Spain—the strategic position of the entire south of France. Occupying the middle between the two extreme points of invasion, it prevents the junction of the two *corps d'armée*, which may have débouched by Bayonne and Perpignan, and enables the French to manœuvre upon both banks of the Garonne as far as the Pyrenees.

12. Bordeaux, chief town of the Department of the Gironde; head-quarters of the twelfth military division; on the left bank of the Garonne, which is crossed by a magnificent bridge one thousand five hundred and ninety-eight feet long; its port can accommodate twelve hundred vessels, each of six hundred tons burden. Two forts, Médoc and Paté, and a citadel, defend it about twelve miles below. Still lower are the forts of Royan and Pointe de Grave.

CHAPTER IV.

TACTICS.

In what it consists—Easy theory—Difficulties of practice—What is required of a
general at the head of a large army—Living providence—Necessity of tactical
talent for subordinate generals—Author at camp of Zeist—Comparisons of strategy with tactics—In what consists genius of war, in regard to both—Skill of
Napoleon—Lützen, sudden battle—Historic detail—Battle not at Starsiedel—
Battle of the Moskowa—Refuse of reserve—Napoleon untrue to his principles
of war—Uselessness of fresh troops the day after battle—Why?—Waterloo—
Guard ordered too late.

Notes.—1. Zeist. 2. Battle of Borodino.

Tactics are the art of handling troops upon a field of battle, and to
make them march without confusion. They treat of the art to maintain order in the midst of apparent disorder, produced by that multitude
of men, horses, and machines, the union of which composes an army,
and to draw the greatest possible advantage from it.

Tactics are the science of the application of manœuvres. A man can
be great in manœuvres without possessing any genius, the perfection
in which is only attained after much practice; nothing is easier to understand than the theory, but the practice is not without difficulties.
The general must be familiar with the means provided and calculated
upon by the orders, that, with a moment's glance of the eye, he should
know how to judge the ground upon which he will be engaged; he must
know how to calculate distances, determine the precise direction, appreciate the details, and to knit the links of circumstances into a chain.

This kind of merit was incomplete in Napoleon, which is explained
by the first portion of his career.

Simply an officer of artillery, until the moment when he was called
to the head of armies, he has never commanded either a regiment,
brigade, division, or *corps d'armée.** He could not have acquired this
faculty of moving troops upon a given ground, which, by varying
incessantly the combinations, developes the skill with every day's move.
The wars of Italy scarcely ever offered to him any application of this
nature, because generally the actions were reduced to conflicts for
posts, the attack and defence of defiles, and to operations in the mountains.

Later, when he had assumed supreme power, the strength of the
armies which he conducted requiring their organization into army

---

* It was Brigadier-General Chanez, commanding at Paris during the winter of
1795–'96, who taught the manœuvres to General Bonaparte, then general-in-chief of
the Army of the Interior.—*Note of Author.*

4*

corps, rendered skill in manœuvres still less necessary. A general, at the head of eighty, one hundred, or one hundred and fifty thousand men, gives but the impulse; he fixes the principal points of movements, issues orders concerning the general circumstances of the army, and provides for the great accidents which occur; he is the living providence of the army. Generals who manœuvre and lead are those who command thirty thousand men and the generals under their orders; they must be familiar with tactics. If I have had some reputation in this respect, I owe it to my long sojourn at the camp of Zeist,[1] where, for more than a year, I have been constantly occupied with the instruction of excellent troops, and sought to instruct myself with that emulation and fervor which a first command-in-chief gives in the beautiful years of youth.

Tactics have the same aim as strategy, but upon a smaller scale and a different theatre. Instead of operating upon a vast country and during entire days, the action is upon a battle-field which the eye can overlook, and where the movements are accomplished in a few hours. The base of combinations, the proposed aim, is always to be stronger than the enemy upon a certain point of the field. The talent is to bring suddenly upon the most accessible and most important positions the means to break the equilibrium and to gain the victory; and finally, to execute, with promptitude, the movements which disconcert the enemy, and which take him unawares.

To effect this, it is essential that the reserves be employed at the right time; and this is the genius of war. It will be carefully avoided to engage them too soon or too late; too soon, is to employ uselessly one's means, and to miss them at the moment when they will be most necessary; too late, is either to permit the victory to remain incomplete, or the reverses to thicken until they become irreparable.

Every one should be obliged to expend the total amount of the energy he possesses; but when the moment of exhaustion comes—and it is that moment which to recognize is so important—then it is urgent to send succor; finally, do not fail to ask for them some time in advance of the urgency.

Napoleon was very skilful in this respect; he knew precisely when the turning point of the battle was at hand. At Lützen he furnished me with a striking example. The battle was begun unexpectedly. Believing that the enemy was retreating, the emperor had left Leipzig with two *corps d'armée*, and had prescribed to me to make a strong *reconnaissance* upon Pegau. Setting out from Wippach, where I had passed the night, I thought it prudent to move by the right of the ravine, although that road was longer; I did not wish to compromise my communications with the main body of the army, which owed its safety to this circumstance. I arrived at Starsiedel, perfectly formed, at the precise moment when the enemy, having surprised the third corps, was about to surround and to destroy it. I had time partly to cover it and to protect its right while it hastened to arms. The battle opened instantly; immense masses of troops, an enormous cavalry, and considerable artillery, attacked me. While the third corps sustained at Kaya a very obstinate conflict with the infantry, Napoleon fell upon that point. The forces which were in my front not ceasing to augment, I sent to him to demand reinforcements; he replied that the battle was at Kaya and not at Starsiedel; and he was right. I had prevented that the battle was not lost in the beginning, but it was gained by him in the centre.

On other occasions Napoleon determined less judiciously.

On the Moskowa[2] he showed a sad want of circumspection in refusing to order his guard to march, although General Belliard asked for it at two o'clock. The Russian army was then in the greatest confusion; immense results would have been obtained with fresh troops; one hour of respite saved the enemy.

Napoleon thus was untrue to one of his favorite principles, which I have heard him pronounce: "Those who retain fresh troops for the day after the battle are nearly always beaten." He added: "If useful the last man should be given, because the day after a complete success there are no other obstacles ahead; prestige alone assures new triumphs to the conqueror."

Alike, at Waterloo, Napoleon gave his guard too late. If it had marched, while the cavalry performed prodigies, the English infantry would probably have been overthrown, and the French army, rid of the English, would have been able to receive, fight, and conquer the Prussians.

In review, tactics can thus be defined: the art of movements, executed in presence of the enemy, with that formation offering most advantages, and which is most in harmony with the circumstances.

NOTES.

1. **Zeist.**—A village in the Netherlands, in the province of Utrecht.

2. **Battle of Borodino.**—This battle, which is also called the Battle of the Moskowa, was fought on the 6th of September, 1812, and opened Napoleon the way into the City of Moscow. The Village of Borodino is seventy-five miles west-southwest of this ancient capital of the Russian empire, and in the centre of an exceedingly strong position, hilly and full of ravines, traversed by the river Kolotscha, an affluent of the Moskowa. In this position, protected by strong redoubts and extensive field-works, the Russians, under Kutusoff, one hundred and thirty-two thousand strong, with six hundred and forty cannon, essayed to bar Napoleon's march upon their capital, who was advancing with part of his forces, quite as

strong, however, as Kutusoff, yet having sixty cannon the less. If an enemy's strength, posted behind such fortifications as the Russians had, and the remains of which can be traced to this very day, is thereby doubled or even tripled, then Napoleon's victory, although one of direct attacks, may be classed among the most remarkable, and certainly the most sanguine he ever fought.

On the day preceding the battle a strong advanced work had been carried by the French with great slaughter. Davoust proposed to Napoleon to march during that night, with forty thousand men, around the extreme left of the Russians, and by Ney simultaneously attacking in the centre, to carry consternation into their ranks. Had Davoust's suggestion been followed, and had the emperor evinced more vigor and resolution individually, there is no doubt that the Russians would have been thoroughly beaten. After a most frightful carnage, the Russians having lost thirty-three generals, fifteen thousand killed, and forty thousand wounded, and the French thirty-five generals, thirteen thousand killed, and thirty-seven thousand wounded, the Russian army remained intact, and slowly retreated upon Moscow. In this battle four hundred pieces of artillery were at one time directed against a single redoubt. Napoleon attacked in *echelons*, with the right under Davoust in front. There Marshals Ney and Davoust earned immortal glory.

CHAPTER V.

MANŒUVRES.

Means of tactics—March and battle order—Formations for both—Deployment mixed with columns—Example—March in the plains of the Tagliamento in 1797—Attack of position—Skirmishers in advance of columns—They cover the deployment —Formation in square—Its speciality—Their difficulties—Squares formed on march—Example in Egypt—Two special causes indicated this formation—Formation of six ranks became superfluous, and is abandoned—Difficulties of formation in squares for march.

Note.—The Tagliamento.

Manœuvres are the means of tactics. They consist in the art of moving masses, and to pass them, without confusion, but with rapidity, from the order of march to the order of battle, even in the midst of fire, and reciprocally.

The battle and the march can be executed with all the formations; but there are preferable formations, both for the battle and the march; and, again, those for the battle vary according to circumstances.

Thus, the deployment is used when the enemy is to be received in position and when he marches, in order to subject him to the greatest fire; otherwise, he would approach scarcely without any danger. If troops march against him, the deployment can still be used, but not without great dangers, by reason of the wavering which the march in line of battle always occasions, and from the disorder which results therefrom. It is then preferable to have only a part of the troops deployed, and to intermix them with columns, which are so many compact points where the authority of the officers experiences less trouble to maintain order. In this formation the right and centre of the French Army of Italy traversed, in 1797, the vast plains of the Tagliamento, and in presence of the Austrian army.

The attack of a position requires the most rapid march, and the ground to be run over being often bristling with obstacles, the troops should always be formed in columns by battalions. These small masses are easily moved; they traverse all defiles without any effort; the rear, less exposed to the fire of the enemy than the head, pushes the latter, and the column arrives sooner.

In order to complete this disposition, numerous skirmishers should precede the columns and march in a direction corresponding with the intervals of battalions, so as to divide the fire of the enemy, and, if necessary, to cover the deployment, without, however, masking the heads of the columns, which can immediately commence firing. Skirmishers thus placed have points of support; their rallying places are designated and within their reach, so that they can never be compromised.

The formation by square can only be accidental, and to resist the attack of a numerous cavalry in an open country. As it is very difficult to move in that formation, especially when engaged with infantry, the troops should be accustomed to pass in the quickest possible manner from the deployed into the deep order, and reciprocally.

Still, in Egypt, we have seen troops marching by squares, and during entire days. But this was owing to two causes: to accustom the soldiers to the impetuous attacks of a new enemy, and to cover the sick, the wounded, and the artillery. The squares were even formed unnecessarily and almost ridiculously heavy, by placing the men in six ranks. It is true that this was suppressed as soon as it became apparent that these precautions were exaggerated, and squares of three and even two ranks were considered sufficient, and then the formation was only employed at the moment when an immediate charge of the enemy could be foreseen.

Generally, the march in square is detestable; however short, it leads to disorders; because the rules of the march are not the same upon the different sides of the square—two march in line of battle, and the others by the flank.

NOTE.

The Tagliamento descends from the mountains which enclose the upper course of the Piave, and taking a course from the north-west to south-east, it washes Tolmézzo, defended by a fort; receives the Fella, which comes from the Gorge of Tarvis, and bathes Chiusa-Veneta, a fortified position; flowing thence, from north to south, it passes Osopo, a fortified place of great importance to the defence of the road from Italy into Austria; forms a multitude of islands and canals; bathes Valvasone, near which, in 1797, Napoleon defeated the Archduke Charles, and finishes its course in the *lagunes*. This river is very important on account of the road which it opens into Germany, and which was that followed by the French in 1797, 1805, and 1809.

PART SECOND

ORGANIZATION, FORMATION, AND MAINTENANCE OF ARMIES.

CHAPTER I.

ORGANIZATION AND FORMATION OF TROOPS.

Both consist of the faculties of the man and the nature of the arm. Order and obedience.

The organization and formation of troops are no arbitrary matters ; their aim being to form a compact mass from an assemblage of men and to make a whole of them, and a unit which is movable; the regulations to be adopted repose upon conditions determined by the faculties of the man, and by the nature of the weapons he uses.

To form troops, order must first be established and obedience be assured. With this aim, a classification and successive bands have been created, which, with skill combined, oblige a large mass of individuals to submit to the action of authority.

FIRST SECTION.

INFANTRY.

In squad, unit is the man—In company, squad forms unit—In battalion, company—In army, battalion—Company for organization and administration—Battalion for manœuvre and battle—True limits of battalion—Conditions of battalion—Necessity to form it with regard to reach of voice—One officer for forty soldiers—Strength of battalions in Austrian and English armies—Inconveniences and advantages of number in the battalion—French battalions—Limit indicated by author—Necessary diminutions of the entire strength—Its greatest reduction when arrived before the enemy—Strength of battalion according to adopted formation—Three ranks—Two ranks—Fire of three ranks—Purely theoretical—Inevitable fusion of third rank into the two first—Cause of disorganization—Object of formation in three ranks—Modifications to be made in formation of two ranks—How this formation becomes best—Formation of regiment—Is arbitrary—Question of administration and economy—Regiments with strong battalions—Their great advantages—Economy, spirit of corps, facility of echelons—Qualities of colonel—Order, justice, and firmness—Special corps —Principles applied—Regiments of light infantry in France and Russia—Only so in name—Chasseurs of Vincennes—Austrian chasseurs—English chasseurs— Voltigeurs—Their application with strong advance guards and in mountain warfare—Necessity of strong companies—Particular instruction, strength, and youth —Good garrison battalions for guard of places—Dangers of confiding the guard to bad corps—The point of support of the army in the field escapes at moment of need—Example : garrison of Ciudad Rodrigo—Major Aubert.

Note.—Remarks on Confederate infantry.

In the beginning a small aggregation, easily governed, was formed ; several of these aggregations were united, and their chiefs submitted to a superior chief; in that case the unit was no longer man, but a union of men.

Thus a squad, composed of eighteen to twenty men, obeys a sergeant, aided by corporals; the squads united form a company, which the captain commands, aided by officers ; and several companies form another mass, which is called battalion. The chief comes into contact with only four, six, or eight men, and he commands through their intervention, and acts thus upon the whole.

The company is the element of organization, discipline, and administration ; the battalion is the true military element in infantry, and the unit for battle ; the movements and manœuvres are by battalion, and by battalion the battle is delivered.

As to strength, the battalion can vary, but in certain limits, determined by the nature of the organizations themselves. The proverb should not be understood literally : *The God of armies is with the heaviest battalions*—a proverb which undoubtedly has been also applied to large armies, designating a part for the whole. Two conditions must be observed in the numerical composition of the battalion : it must be movable ; and when deployed, the voice which commands must be heard at the two extremities of the line. Within these limits the number of companies composing a battalion, and the strength of each, can more or less be increased.

A proportion must be established between the number of officers and soldiers. That indicated by experience as agreeing best with the economy of a well-established service is one officer for forty soldiers, or twenty-five officers for a battalion of one thousand men. It must be, however, understood that a large number of officers has but one inconvenience—that of costing too much to the state ; in every other respect it is useful, be it in multiplying the means of action, of surveillance, and of examples of courage, or in facilitating rewards by a more rapid advancement.

The effective of the organization varies with the different nations. The strongest battalions are in Austria; England has the weakest.

The total in Austria exceeds twelve hundred men; this is too much for a good service; such a number can not be moved with order and facility.

I see, however, an advantage in this disposition; since the losses during war continually take place, and when reinforcements have not yet arrived, a battalion of such strength can resist longer; a large dimi. nution of strength does not unfit it for service.

In France weak battalions are habitually employed, and their effective strength, even upon the entrance of a campaign, is still almost always below the complement of organization.

I shall put one thousand men as the limit for the strength of a battalion, because even this number is not always maintained during peace, and at the moment when it leaves the garrison to take the field. After constant observation I have found that the best administered battalion is the strongest one; and assuming a diminution of one-fifth for hospital details, workmen who remain with the depots, those required for the trains, etc., etc., a battalion of one thousand men will then have but eight hundred men under arms; after several months of active service it is reduced to five hundred—a force still sufficient to face the enemy.

The formation adopted for battalions influences their numerical composition also.

In all armies of the Continent, infantry is formed in three ranks; it is formed in two in England. This last formation appears to me far preferable. Nothing justifies a third rank.

Without entering into the detail of fires, I will appeal to experience. On drill the fire may be delivered in three ranks, but not in war. The French regulations prescribe that the weapon be passed to the third rank, which is uniformly required to load it. This theory is not applicable before the enemy, and long practice has demonstrated its inutility. The battle is fought by firing when in position. The best formation, then, is that which mostly facilitates firing, gives it a better direction, and the greatest development; in fact, the third rank will soon be confounded with the two first, because instinct teaches the most advantageous formation; but this change being against orders, it brings about a kind of disorganization; it is, therefore, better to sanction this formation at once, and make it permanent.

In placing troops in three ranks, the object has undoubtedly been to give greater consistence to the march in line of battle; but this means is not sufficient. Even with three ranks, a line moving is but little solid; and for the march in line of battle I should prefer a less heavy formation.

5

With a slight modification, the formation in two ranks is equal to all requirements. Thus :

In position, the troops, when formed in two ranks, have for action a front by one-half greater than they would have were they formed in three ranks. In the march in line of battle, ploy the first and fourth divisions in rear of the second and third, and you will have four ranks; and at the moment of halting you will present a front, it is true, less by one-fifth than that of the actual formation (four divisions in line of battle, each composed of three ranks), but in two minutes it will be doubled. Here is, then, a solid and compact formation for the march, which permits a battalion to fire everywhere in case of a sudden charge of the surrounding cavalry of the enemy, by a simple about-face, executed by the first and fourth divisions, which double the second and third.

The formation in two ranks, with this disposition introduced into the march by line of battle, appears to me incontestably the best.

After the formation of the battalion comes that of the regiment. Here everything is arbitrary, and depends upon the caprices of those in power ; the regiment may be two or three, or from four to five and six battalions strong ; it is only a question of administration and economy. numbers of men ; there are less staff officers, and the advantages arising Regiments composed of many battalions are less costly, with equal from living together are given to a greater number of men. Generally, these regiments are in better moral condition and have a more energetic *esprit de corps*, because a larger number of individuals partake of its reputation and glory. The consciousness of their ability to distinguish themselves is more powerfully developed—so much so, that they would cheerfully endeavor to execute the most heroic deeds. In wars of invasion, in the occupation of vast countries, regiments thus constituted are able to form echelons to assemble the men who have remained in the rear. These intervening corps receive the recruits, drill them, and strengthen the battalions which are in face of the enemy. Thus a great economy in men is obtained—an economy which is not less important than that of money.

In general, the regiment is an essentially administrative formation; it again acquires a kind of social constitution, animated by a love equal to that of country and of home.

The colonel is the chief of this social assembly—its father and magistrate ; and certainly, without wishing to depreciate his courage, the first military virtue, the essential qualities of a colonel, and those which chiefly influence the efficiency of his regiment, are less an extraordinary intrepidity, than the spirit of order, of justice, and great firmness. The best corps are those thus commanded.

It should be a principle to instruct a regiment of infantry in all branches of the service, and the exigencies of war point to the light infantry as the most proper organization. However, special corps have been considered useful, and I partake of this opinion. For advanced guards, and detachments in broken and mountain countries, men are required who have been endowed with a particular instruction—who know, by instinct, how to surmount all obstacles—and who, trained with the greatest address, can make their fire terribly murderous.

But, as far as I am informed, the true principles have not been applied in every country.

In France and Russia there are regiments of light infantry; these corps differ from ordinary regiments of the line only in name and uniform.

The Chasseurs de Vincennes have recently been established in France. The institution is good, but incomplete, as long as the battalions which compose this corps are not divided into field and garrison battalions, according to the principles I hereafter shall lay down.

In Austria there are battalions of chasseurs; in England, companies, belonging to a regiment, which never leave the depot. These two organizations are superior to ours; but they stand yet in need of being modified.

The regiments of infantry have their voltigeurs; in this respect an immediate necessity is satisfied. In recruiting the voltigeurs of the centre companies, men can always be chosen capable of performing good service.

Special corps of light infantry should have a force proportionate to the necessities of heavy vanguards and mountain warfare. Regiments of several battalions are too strong for this service; and since it necessitates an extremely great division of the men, a chief can only command a small number. An organization should, then, be adopted which presents to the enemy only a strong battalion.

In this case the companies should be very strong. I should wish to see a battalion of light infantry twelve hundred men strong, composed of six companies of two hundred men each, commanded by five officers. But the particular instruction and the special formation of these troops is not the only thing required; they should have more strength, and should be younger than other troops; the choice of the men is most important.

If a new corps is formed, it can be constituted in the most satisfactory manner; but, at the expiration of several years, there will be found, for the training and perfection of young soldiers, nothing but heavily moving skeletons, and the corps will have lost all its former agility.

Light infantry corps should be composed of two battalions—one of twelve hundred men, to be always maintained at its full complement and constantly ready for war; the other of four companies, composed of six or eight hundred men, which I will call garrison battalion, designed to instruct the recruits, and to receive all men still fit for service, but who are no longer suitable for outpost duty, which requires so much strength and youth.

I see another advantage in this disposition; excellent troops are placed at the disposal of a general, with which he can either garrison fortresses or protect fortified posts menaced by the enemy.

I know that the resolution is very hard to place in a fortress a good regiment, or a part of one, which is able to go into the field; but, nevertheless, it is absurd and sad to confide the guard and the defence to bad corps. They surrender the place at the first attacks of the enemy, and the general sees the point of support upon which he counted vanish at the moment when most necessary to him.

I have twice painfully experienced this in Spain. General Dorsenne had formed the garrison of Ciudad Rodrigo with negligence; and this fortress, which had held out for twenty-five days of opened trenches against the French army and the most powerful material, was taken in four days by the English, while the Army of Portugal was hastening to its succor.

Later, I had with the greatest care fortified the passage of the Tagus at Almaras, in order to render sure the communication of the Army of Portugal with that of Southern Spain. Works strengthened with masonry and defended by small redoubts covered the left bank, and advanced forts disputed the only passages by which the enemy's artillery could debouch. This post of Almaras was of great importance, and I had placed in it garrisons of sufficient strength. But these troops were mixed, and the bad ones had the majority—especially a German battalion, called *Prussian*. The good troops occupied the outposts defending the Pass of Miravette. The enemy appeared suddenly; the English column, with which the artillery marched, stopped, and could not pass. But another column, having traversed by foot-paths the girdle of rocks which border the plateau, arrived with ladders, and mounted to the assault. The slightest resistance was sufficient to repulse an attack so audacious, executed in full daylight. The commandant of the fort, Major Aubert, a very brave soldier, mounted the parapet to encourage his intimidated troops; he was killed; his death spread terror among his men, and the garrison fled to the other bank of the Tagus, abandoning the fort to the enemy, who retired after having destroyed the means of defence.

NOTES.

Remarks on Confederate Infantry.—I believe our system of regiments with ten companies to be unwieldy; and that of two battalions, each of six hundred and forty men, divided into *four* companies, with *sixteen* line-officers, and commanded by the lieutenant-colonel and major respectively, the whole to be directed by the colonel, to be much more effective; because:

1. Six hundred and forty men, in but four companies, will move better, more quickly; and straggling, especially in long marches, will be less frequent.

2. The lieutenant-colonel and major would take more interest in their respective commands, since they will have the consciousness of being in command in reality and always. They would be thus forced to take greater interest, even if they did not have the desire. As now, the duties of these field-officers are not sufficiently defined. It is true, in battle they have charge of the right and left wings respectively, but in many cases this appears the entire performance of their position.

Colonels are not always willing to divide their important duties with officers in whom they do not repose entire confidence. Such a thing may not happen in regular armies, established for centuries, but it is of daily occurrence in an army called into existence as ours has been, and where reliance in an officer's desire for improvement and strict performance of his duties had to take the place of the requirements necessary for his official position, which it was supposed he would be anxious to secure. If it is voluntary on the part of a colonel to assign such duties to his field-officers as he may think fit, he will often only yield to the demands of necessity, whenever he is so overworked as to make assistance absolutely requisite— from no other motive, but that what he himself does he knows to be done properly.

But if, on the other hand, the organization of an army is such as to give to the two subordinate field-officers the powers of the colonel, as now possessed, their sense of duty will be stimulated; because they then will, in reality, have specific services to perform, the neglect of which they know will subject them to ignominy and punishment.

3. Then the colonel's position would be more important and becoming to his rank. He would no longer be the commanding officer of a battalion, but that of a regiment consisting of two battalions; performing the duties, on a smaller scale, of a general of brigade; and better fitted to assume the latter command, as colonels now generally are, whenever the casualties of war call them to some temporary or permanent command of the kind. A colonel, moreover, no longer fettered by small duties of every kind, would have more time left for the perfection of his regiment, and the majority of brigade manœuvres could be performed by his regiment at any time.

4. One line-officer to every thirty-one men appears more than sufficient. An officer can easily manage forty men.

5. Whenever a portion of such a regiment would be required for separate duty, it would still feel itself an efficient body, and have reliance as such, since it would still preserve a regular organization, having another battalion around which to rally, which is not the case if four companies are taken away from ten. For this reason independent battalions of less than ten companies can not be effective. This is a matter worthy of consideration, especially on account of the requirements of the field of battle.

6. Brigades could then be efficiently regulated; and a brigadier ought never to command more than three of such regiments, which would give to him a front of battle of six battalions, or two battalions for each division (right, centre, left) of the line, consisting in all of three thousand eight hundred and forty men under arms.

7. It would reduce the number of officers; an important consideration, whenever it is admitted that it be the duty of a country to provide for those officers who become disabled in the service.

5*

8. A beneficial emulation would be excited between the two battalions of the regiment.

If mobility, *corps d'esprit*, and economy are important considerations, I believe they are attained by a like organization. The following conditions, however, are inseparable:

1. The battalions must never be permitted to fall below an effective of two-thirds, that is, four hundred men under arms. Below that number a downward course commences which nothing can check. To gain this point, the distribution of recruits should be more just than it has been in many cases. The general commanding alone, it appears to me, could equitably determine which regiments may be entitled to the first recruits—for it must be owned there is a difference between regiments which have become depleted through the bad management of their colonels and insufficient attention of their sanitary officers, or those whose *corps d'esprit* is not sufficiently established to prevent desertions *en force*, and those whose thin ranks are to be ascribed to their valor upon the field of battle.

To keep up regiments in a proper strength is so important that it deserves the most serious attention. I believe that the bravest man will fight better when he knows he can, in case of need, rely upon some succor. Soldiers of those regiments will fight best who can yet boast of a becoming strength of their commands, and when they are conscious that they will not be permitted to waste themselves away without reinforcements; that, on the contrary, they will be upheld as long as their country has yet a name. Those brigades will fight best who, by reason of equal dimensions of their commands, have the additional weight which superior mobility, impulsion, and spirit confers.

Justice demands that they be upheld in strong numbers, because, in the eyes of the world, and, alas! in those of many commanders, a regiment is a regiment, and expected to perform equally heavy duties with the strongest. The pride of these glorious old regiments will never lead them to complain of hardships. They have learned, upon many a hard-fought field, that it is valor, combined with obedience, which gives victory, and that the country's independence is never won by complaints. Their bravery will be the same when marching to death with but one hundred and twenty-five men, conscious that not one-half of them will issue from the smoke of battle; but it is rather the bravery of despair—like that of the Last Ten of the Fourth regiment of Poland.

If we look at the percentage of the killed and wounded of the different regiments engaged in a given battle, we will find that where *small* regiments lose fifty per cent. of their number, *strong* bodies pass the ordeal with a loss of but twenty-five per cent. To properly estimate their services upon the field of battle, the losses of regiments must be compared with their previous effective strength.

This appears, beyond controversy, to be established through the consideration of the *killed* and *wounded* during the Battle of Shiloh, the materials for which I have taken from the "Official Reports of Battles, published by order of Congress, in 1862."

> General Bragg had..13,589 men
> Lieutenant-General Polk had...................... .... 9,136 "
> Lieutenant-General Hardee had .... 6,789 "
> Major-General Breckinridge had...................................... 6,439 "

And, in the ratio of their effective strength, it appears that the *percentage of killed and wounded increases with smaller corps.*

These percentages of loss are:

> Bragg's corps lost ....22.04 per cent.

Polk's corps lost ....25.60 per cent.
Hardee's corps lost...30.03 "   "
Breckinridge's corps lost ....32.28 "   "

or precisely *ten men more from every one hundred men than did General Bragg, who commanded the strongest corps.*

Gardner's cavalry, in killed, wounded, and *missing*—which latter are not included in the previous calculation—lost 6.86 per cent., but they were hardly engaged.

The Battle of Shiloh furnishes a very fine example, because the troops were inspired with the same degree of ardor, and the different corps were equally well commanded. A more uniform battle, in almost every respect, can hardly be conceived. The brigades of Cleburne and of the gallant Statham—in which latter my regiment had the honor to serve—lost, it appears from the same report, heavier than any others upon the field—the first nine hundred and seventy-eight men, and the latter seven hundred and sixty-four men. Statham's brigade was one of the smallest, barely numbering two thousand men.

2. It is a question to be determined by the Department of War how long it may be expedient to permit conscripts, especially those who have withdrawn their service from the cause of the country so long, to volunteer. After a certain indulgence this boon no longer belongs to them.

3. A thorough system of consolidation, based upon the just foundation of the *corps d'esprit* displayed by the different regiments and their exploits, can alone save the efficiency of the army, and to carry out any good system of reorganization it is absolutely necessary. The unanimous and spontaneous re-enlistments of our noble armies in the field show that no sacrifice will be too great for such men. Soldiers are the very first to see the necessity of reform. The following basis for consolidation appears to me a just one:

Let those regiments be preserved *first*, who enlisted *for the war* at the beginning of the struggle.

Second. Those who enlisted for *three years.*

Third. Those who *unanimously* re-enlisted, according to General Orders No. 1, Adjutant and Inspector-General's office, January 1, 1862.

Fourth. Regiments established from such fragments of one year regiments as declined to re-enlist entire.

Fifth and *last.* Those regiments which were organized afterward.

Under the restriction, however, that each and every regiment which shall have been found to be wanting in *corps d'esprit* and bravery be disbanded.

Such a basis rests upon good conduct and patriotism, and no other consideration can be admitted.

4. We have, then, to increase the number of non-commissioned officers. Upon their efficiency the discipline of a regiment mainly depends.

5. Each battalion must then have a separate non-commissioned staff. The commissioned staff, as now existing, is deemed sufficient, and, if necessary, two subalterns of the line may be detailed in rotation to the field-officers commanding battalions, to act as adjutants.

6. The number of officers of the line and field must then be continually at its full complement. To effect this, the establishment of a reserve list is the most effective measure. As soon as the absence of an officer, from any legitimate cause, exceeds a certain period, disorganization commences; because, and especially in our own service, what is called "electioneering," a remnant of political practices, begins— a practice which sullies the honor of every officer engaged therein. Hence it is important that their places be filled at once. This practice arose from the volun-

teer system, but it is fast disappearing before the rigors of executive justice, military honor, and discipline.

SECOND SECTION.

CAVALRY.[1]

Fire-arms are accessory—Its speciality in the combat of corps by corps—Impetuosity of its movements—Superiority of French cavalry—Its exploits in Italy—Scarcity of good commanders—Three counted in twenty years of warfare—Their names—Indispensable qualities of a cavalry general—Promptitude in surveying condition of affairs—Rapidity of decision—Prodigality and dash in the attack—Conservation and minute attention to needs before battle—Employ without reserve at the given moment—Particular character of Murat—The march of column without object before the enemy—Why—Necessity of formation in two ranks—Squadron unit of combat—Mobility combined with regularity; basis of its strength—Best reputed formation—Unit of combat same as unit of adminis tration—Faulty organization under the Empire, reformed at peace—Line of cavalry in battles; divers formations of the French, Germans, and Russians—Part of cavalry—Sees and hears for the army—Gains the victory—Lützen and Bautzen—Incomplete victories—Cavalry in battle—Two objects to accomplish—Heavy cavalry—Lances and sabres—Light cavalry—Fire-arms, carbines, and pistols—Dragoons—Mounted infantry—Speciality of that arm—Beliefs and convictions—Necessity to employ small horses—Shoeing and clothing—Armament of cavalry in general—Lances, sabres, pistols, and carbines—Lance in particular—Little suited for light cavalry—Its origin—True arm of cavalry of the line—Sabre and lance compared—Lance at Dresden—Cuirassiers repulsed—Breach made by fifty lances—Prejudices of routine—Cossacks—Their aptitude—Formed by nature—Cossacks of the Don—Degenerated—Cossacks of Asia, of Kouban—Russian hussars and chasseurs, cavalry of the line—How Austria might have Cossacks—France will have hers—How—Defensive arms as suited for cavalry—The best—Should be also applied to infantry—Instruction of cavalry—Its true object—Horse and rider—Equitation—Accustomed to charge—Promptly rallied—Inconvenience of light charges when instructed—Sad education of the animal—New indications given by the author.

Notes.—1. Remarks on Confederate cavalry. 2. Marshal Murat. 3. General Nansouty.

In cavalry as in infantry, above all, the object should be the attainment of order, obedience, and mobility; but the manner of combating and the nature of weapons not being the same, everything is different in the application of that arm.

Fire-arms are with cavalry an almost superfluous accessory; most frequently they only serve as means for signals.

Cavalry is destined to fight body to body; it must cross swords with the enemy, dash at him, overthrow, and pursue him. The pursuit of the enemy is its habitual office; for but rarely two parties engage in a hand-to-hand encounter. At the moment of their meeting the less confident one stops and flies.

The movements of this arm must always be rapid and impetuous ; sometimes even, but with small bodies only, it should fight with a hardihood bordering on imprudence.

For the conflict the French cavalry is the best in the world ; it always charges with the utmost impetuosity.   It will sometimes become the victim of rash enterprises ; but, in general, what favorable results are secured through this habit of headlong boldness !   In our first immortal campaigns of Italy, how many thousands of prisoners we owed to a handful of horsemen !

To command cavalry, and to direct considerable masses, superior qualities and particular merit are required.   Nothing is more rarely met with than a man who knows how to handle and to conduct these masses so as to employ them at the right time and in the proper manner.   But three can be enumerated in the French armies during a period of twenty years of warfare : Kellermann, Montbrun, and Lasalle.

The necessary qualities of a general of cavalry are of a nature so varied and so rarely united in the same person, that they almost appear an isolated, particular gift.

Above all, the talent is required to survey, with one glance of the eye, the situation, surely and promptly ; a rapid and energetic decision, which, however, does not exclude prudence, is next; because an error or a fault committed at the beginning of a movement, are irreparable, by reason of the limited time which remains for execution.   It is otherwise with infantry, the march of which, compared with that of a general and his aides-de-camp, is always slow.

The general of cavalry should study to shelter his troops from the fire of the enemy as long as they are in position ; but when the moment of launching them against the enemy arrives, he should be prodigal. Upon the eve of a battle, and until they are called upon to engage the enemy, he should administer to the comfort of men and horses with the minutest care ; he should keep his forces in the highest condition, both physically and morally ; but, at the arrival of the proper moment, he should expend this cavalry without regard to the chances of loss, with the sole purpose of achieving the greatest advantage.

A general scarcely ever satisfies, in the same degree, these two requirements.   One, an excellent administrator, preserves his cavalry ; but, too much occupied with this thought, he does not dare to launch it against the enemy, and when the day of battle comes it is useless. Another, always ready to go into action, takes so little care of it during the campaign, that his troops perish miserably before they see an enemy.   To cite two examples: This defect of careful treatment will be charged upon Murat,[2] and the contrary excess upon General ———[3], who commanded the cavalry of the Imperial guard at Wagram after Bessières was wounded.   If he had charged at the moment when

Macdonald made his offensive movement, sustained by the artillery of the Imperial guard, and when he had overthrown the Austrian right, twenty thousand prisoners would have fallen into our hands.

Before the cavalry charges the enemy and the men fight against each other, it should never attack in columns. This formation will serve to facilitate the march, but at the instant the enemy is approached the deployment should begin. A column of cavalry, once surrounded, is soon destroyed, because there are but few soldiers who then can make use of their arms. The cavalry should deploy in two ranks, so as to arrest the disorders which happen in the first—formerly it was deployed in three; it will not be difficult to recognize the evil of this formation.

The unit of combat is the squadron; the rule to determine its strength is to combine the greatest mobility with maintenance of order.

A squadron having too large a front will be easily put into disorder by the smallest obstacle, and every troop in disorder is half conquered. According to experience, the best formation, and that which joins the greatest strength and consistence with great mobility, is a squadron of forty-eight files, divided into four platoons of twelve each. Platoons of sixteen and eighteen files are also suitable at the beginning of a campaign, especially of light troops, when more active service and numerous detachments weaken the corps.

The less number of men and horses permits that in cavalry which would be impossible in infantry, namely: the unit of combat is the same as the unit of administration.

The perfection of the service, generally, would demand such an organization of all the arms as could both be applied to the combat and the daily existence—that is, to the police of barracks, for the purposes of administration and manœuvres; an organization which constantly keeps the troops under the hands of the same chiefs, and which thus would impart greater stability and power.

Formerly squadrons were composed of two companies. One of the captains found himself subordinate to the other—a vicious combination. He who commands must have a social superiority, constant and determined, over those who obey; such is the fundamental principle of military subordination. Still we have made war with squadrons thus formed; but, since peace, a profound discussion has united all factions; the company-squadron has been adopted, and the soldiers, whatever their position and the circumstances, are governed by one chief.*

---

*Lieutenant-General Préval, who, under the Restoration, was member of the Council of War, and one of its luminaries, is the author of the principal ameliorations which have been brought about in the organization of the cavalry.—*Note of Author.*

There is a difference in the manner of the French, German, and Russian armies, as regards the formation of the cavalry in line of battle. With us the squadrons have regular intervals; with those foreign armies they are joined by twos, and form a division without intervals. This formation, while preserving the same degree of mobility in the squadrons, gives more consistence to each point of the line, and, in that respect, it has some advantage; but, on the other hand, with the French formation, there is a line, the front of which is larger, with an equal number of combatants, which gives a greater expansion to the wings. I will not decide in regard to these two formations, the advantages and drawbacks of which appear to me equally balanced.

Cavalry is necessary in war to disclose and give news of the enemy. Such is, especially, the duty of the light cavalry; it is the eye and the ear of the army; without it a general is always surrounded with peril.

Cavalry is, furthermore, useful for combat and to profit from victory. Without cavalry, a gained battle gives no decisive result.

A proof of this can be cited. In 1813, after having beaten the Russians and Prussians at Lützen and Bautzen with infantry alone, as far as the moral condition of the army was concerned these victories were of great importance, but no real advantages resulted from them. A flying enemy can always rally, if he is not rapidly followed up at the moment of disorder.

Cavalry in battle has a twofold object: 1. To engage the enemy's cavalry and to pursue the beaten army. 2. To engage the infantry which may throw itself in the way.

To engage infantry, heavy and iron-clad cavalry is necessary, which is sufficiently protected and sheltered from the fire, so as to confront it fearlessly. It should be armed with lances and sabres; each man will be armed with a pistol; no other fire-arms are necessary except a certain number of carbines per squadron, so that each regiment have the means of clearing itself, should it be isolated.

There is a fourth kind of mounted troops, a very old institution,* and which has been misapplied—it is not known why; I speak of the dragoons.

In principle they were entirely mounted infantry; they ought always to have preserved this character. In that capacity dragoons can render, in a thousand circumstances, immense services—such as detachments for surprises in retrograde movements, and, above all, in pursuits. But it would be necessary, in conformity to their institution, to mount them upon horses too small to be placed in line; otherwise the aspirations and pretensions of colonels will soon convert them into cavalry, and they will then become both bad infantry and cavalry.

---

*Marshal Brissac, in the 15th century, during the Piedmontese wars, raised the first corps of dragoons, and used it to great advantage.—*Note of Author.*

A body of troops should have its beliefs, its convictions, and its faith, resulting from sanctioned principles, and even prejudices, inculcated in the minds. But the intelligence of the soldiers must not be confused by the profession of different opinions; so as to say, for instance, in a solemn manner, when exercising them on horseback, that cavalry should always triumph over infantry; and when the moment of exercises on foot arrives, to teach them *per contra*, how good infantry is invincible by cavalry. With their application the axioms recur to the minds of the soldier, and almost generally in a reverse manner. As a foot-soldier he recollects how redoubtable the cavalry is; as a horseman, he never forgets how much the infantry is to be dreaded by the cavalry.

I repeat, that no institution is more useful than that of the dragoons; but it must not be falsely employed. Let the horses be small, as already said; let their harness and equipment of men and horses be uniformly calculated for the commodious and rapid service of a true infantry corps, armed with good guns and bayonets, and well supplied with ammunition. Let the dragoons, lastly, be shod and clad for easy and rapid marches.

As for the regular cavalry, cavalry of the line, and cuirassiers, I would compose their armament of lances and half-curved sabres, suited for the double purpose of cutting and thrusting, and of a pistol; each squadron should have twenty breech-loading carbines.

In another work I have considered the question of the lance. Not to leave the matter of which I am speaking incomplete, I will here reproduce the arguments recommending that weapon, according to the expression of the Marshal of Saxe: *The queen of arms.*

I will, then, commence at once by remarking that it is totally unfit for light cavalry, which, having to defend itself against several enemies at the same time, should be provided with fire-arms and sabres. Yet the light infantry has been armed with the lance in those countries in which it was introduced.

But it is known with what facility new usages are adopted; in the most civilized countries the authority of example leads to a blind confidence. The origin is never traced, nor the circumstances which explain it; essential differences are not taken into account, and from this dates the beginning of faulty and inconsiderate applications.

From whence comes, then, the false employment of the lance in the arming of mounted troops? From the example given by warlike hordes, such as the Cossacks and Arabs. These hordes inhabit plains where horses are abundant; they fight without instruction and rules, and employ the lance in a wonderful manner. Therefore it has been said, by considering them light troops, the lance should be of service to light cavalry.

And neither has it been sought to ascertain the origin of this weapon, nor why these hordes employ it so skilfully.

In a barbarous country, where no kind of industry has, as yet, penetrated, where neither manufactories nor magazines of arms exist, nor money to buy the latter abroad, a man mounts a horse and looks about for a weapon; he cuts a long limb of light wood, sharpens a point, hardens it near the fire, and he has a lance. Afterward he procures a nail, puts it in, and his weapon is more dangerous. Finally, this staff is clad with iron regularly fashioned, and he has a lance such as has been adopted for troops.

The Cossacks and Arabs did not arm themselves in such a manner by choice, but by necessity; and if they have become redoubtable for their skill in handling the lance, it is because they have exercised with it from infancy.

Nothing, then, can be drawn from such examples for light troops, specially organized in a civilized country.

The lance is the weapon of the cavalry of the line, and principally destined to combat infantry. The sabre can not take its place; what use will cavalry make of sabres if the infantry remains firm, and is not afraid? The horsemen can not sabre the foot-soldier, because the bayonets keep the horse at too great a distance. On the other hand, if the horse, which remains the only offensive arm of the horseman, be killed, it falls and makes a breach, and this breach gives to those who are near it the means to penetrate. On the whole, the advantage of the struggle then is with the infantry. On the contrary, suppose the same line of cavalry, furnished with a range of pikes preceding the horses by four feet, and the chances of success will be entirely different.

But for light troops the sabre is better suited than the lance; in hand-to-hand conflicts a short weapon is managed with more facility, and is more advantageous, than a long one. All things being equal, it is certain that a hussar or a chasseur will beat a lancer; they have the time to parry and thrust again before the lancer, who, beset by them, will be obliged anew to resume the defensive.

The sabre, destined for light troops, should be slightly curved; the perfectly straight sabre is less suited for single combat.

The same troops should, in addition, be supplied with fire-arms, either to increase their power of resistance or to be able to announce their approach to troops which they are ordered to warn or to sustain.

As regards cuirassiers and the entire cavalry of the line, it would be suitable that they have both the lance and the straight sabre. The first rank would charge with the lance; the second sabre in hand. The shock once produced and the ranks broken, the sabres of the second rank will complete their work.

In times of chivalry the conflict was always by the front, and the stroke came direct; the long weapon then would have been preferable. This explains to us the use of the lance by the knights.

In support of my opinion, I will cite a fact in regard to the manner of employing the lance and obtaining great effects from it.

In 1813, at the Battle of Dresden, our cuirassiers had several times charged the infantry, abandoned by the cavalry, upon the left of the Austrian army. The infantry always resisted; it repulsed our attacks, although the rain had prevented almost all the muskets being fired.

This resistance was not overcome until the cuirassiers were preceded by fifty lancers, forming the escort of General La Tour Maubourg. The lancers made a breach; the cuirassiers were able to penetrate, and destroyed all. It is true the infantry fired but few shots; but under every other circumstance the question would not have been uncertain had the cuirassiers been armed with the redoubtable lance.

The lance is equally victorious in cavalry engagements, line against line, and when the enemy has only sabres. It is admirable at the moment of attack; and it is not less suitable for pursuit.

To resume. I am, then, authorized in saying that the lance should be the principal weapon of the cavalry of the line, and the sabre only an auxiliary one; and that the armament of light troops should consist in sabres and fire-arms. Undoubtedly, routine and contrary prejudices will yet, for a long time, combat these principles, whose truth seems to me, however, to be perfectly demonstrated.

The Russian army has an immense advantage over all other European armies. The Cossacks serving in it compose a light, admirable, indefatigable, and intelligent cavalry; they know with precision what to do, and where they are; they reconnoitre the country well, observe everything, and can perfectly rely upon themselves. They can not be compared with any systematically instructed light cavalry; nature has formed them for this service; their intelligence is developed through the daily necessities to which they are subjected. I speak of the Cosacks of the frontiers, who, continually at war with their neighbors, always in the presence of an enterprising and tricky enemy, are obliged to watch unceasingly for their own security.

The Cossacks of the Don, who were formerly admirable, have become less efficient and intelligent since their country is surrounded by subjugated provinces. Still, numerous tribes of Cossacks remain to guard the frontiers of Asia, especially those on the Kouban, upon the line, and Terek, and to the east of the Caspian sea. During a war, Russia can dispose of and bring to Germany more than fifty thousand of this cavalry, which leaves the regular cavalry in careful preservation for the day of battle. This circumstance permits Russian hussars and chasseurs to be considered cavalry of the line, and prevents them from being employed as light troops; because, on account of their different instruction and non-employment as light troops, they understand scarcely anything of those duties which the Cossacks perform so admirably.

Austria could have something analogous to the Cossacks, but not on so large a scale. She could easily procure ten thousand troops of this kind by forming a corps of five hundred horse in each frontier regiment. I do not understand why, in a country where everything is combined with such care, and where the organizations are so well considered, something resembling it has not, as yet, been introduced.

France, when she shall have conquered Algeria, can without any trouble levy Arab troops which, in times of war, will render her incalculable service. The attainment of this object calls for the unremitting attention of the government; and to arrive at this end, it would be well at once to increase the native troops as much as possible, so as to have a host of men attached to the glory of our arms, accustomed to unite their interests with ours, to rejoice with our success, and to be able to furnish good non-commissioned officers, whose want will be felt in proportion as the organization extends itself.

The cavalry being destined to fight body against body, it may be asked, why the question has not been considered, to secure it from the blows of the enemy? Little would suffice to guard it from a sabre cut, a thrust of the lance, and even to deaden a musket ball, fired from a short distance, or a pistol-shot. The Orientals, whose conflicts are always mêlées, have had this foresight at all times, and are often clad with mailed coats. The bust could be guarded by a coat of mail of buffalo-skin, such as the Castilian peasants wear; as for the head, the shako could be lined within by two crossed pieces of wood, as it is sometimes used; the limbs should be protected by one or two light iron plates, placed on the outside of the sleeves and pantaloons. This double cuirass of buffalo-skin, trimmed and ornamented, would make an elegant dress, resembling that of the Roman soldiers; and this light and comfortable attire might, perhaps, be equally suited for infantry of the line, and which, favorable to health, would protect the soldiers from the grievous effects of a change of clothing to suit the temperature. The dress would then be reduced to a sort of vest, like that of the cuirassiers; and the dress of buffalo-skin, worn only when under arms, would be the insignia of service.

I will add one word concerning the instruction of the cavalry, which has always appeared to me to be incomplete. Too much attention can never be bestowed upon equitation, and to make the horseman a perfect master of his animal. Man and horse should make but one individual— thus realizing the centaurs of fabulous memory.

Horsemanship is everything. It subjugates and breaks the horse. Manœuvres will always be performed with sufficient correctness when the soldiers are good horsemen. Encouragement of all kind should be given to obtain this result. The troops must, furthermore, be instructed to charge thoroughly, without occupying themselves particularly with preserving order, which this impetuous manner of the movement would render impossible, and which is the best means to beat the enemy; but, at the same time, they must be accustomed to rally at the first signal with promptness and dexterity. They must be constantly reminded of this, and prepared for it with all means. The apparent disorder of the charge will not then influence their moral state.

On the contrary, if the charges are made moderate in instruction, being in consequence, still less lively before the enemy, they will never overthrow him; at the first disorder the soldiers will believe themselves lost. But instructed, as I have before said, they will consider this disorder an habitual circumstance, easily repaired, and without any danger.

The following mode is often employed in drill, in great manœuvres, and in sham-battles : The infantry is charged by the cavalry, and, bearing in mind that it is only a sham-fight, the cavalry stops before having reached the infantry, or escapes through the intervals. Nothing is worse than such an education of horses; by thus accustoming them to avoid obstacles, they will never be brought to confront them, because their habits will accord with their instinct, and, perhaps, with that of the rider's. This practice is pernicious; it should be banished from the exercises, and replaced by perfectly opposite lessons. The results will be immense in war. I conceive it in the following manner :

A line of infantry is placed to face a line of cavalry; the files of both lines are so divided that a horse and man can easily pass through the intervals. At first, the cavalry moves off at a pace, and rides through the infantry; this is repeated several times at the trot and gallop, until the horses execute this movement, so to speak, by themselves. It is then accompanied by several shots, which fire is extended along the entire line, and which gradually becomes heavier; and, if it be wished that the noise be still greater, files of infantry of six ranks may be formed, and the actual shock of the fire of an entire battalion is produced.

After several days of like exercise, such cavalry will be more suited to attack infantry than any other; and the horses being drilled and accustomed to throw themselves upon heavy discharges in their very front, they will carry their riders by themselves, even if the latter were endeavoring to moderate their ardor.

NOTES.

1. **Remarks on Confederate Cavalry.**—That we, with such magnificent material, should see our cavalry occupying the third place among the three arms, in both efficiency or usefulness, must be owing either to some radical defect in our system of organization, or to our want of experienced cavalry officers—to which, in particular, I do not feel prepared to decide.

In all modern wars cavalry has played a great *rôle*. It is the only weapon which can take advantage of a beaten enemy, pursue, and rout him. Hence it has been fostered and trained in every well-regulated military system. Frederic the Great owes his most signal victories to the impetuous attacks of his hussars. Napoleon hurled his masses of heavy cavalry against the decisive point, and felt assured of victory. The Polish lancers dashed into many a Russian square in 1830 and 1831. The cavalry charges of Inkerman and Balaklava are immortal. On the plains of Italy both French and Austrian horsemen, in the late war, performed prodigies. And all these bright deeds of arms were performed by men who had to be taught to ride *after they were twenty years of age, and upon horses that had been broken by riding-masters who themselves would not have dared to mount a wild young horse, as our cavalrists do.*

There, severe application gives, in a small degree, what our national customs richly bestow upon us; and yet, what immense disparity in the achievements!

The fact is, that we content ourselves with proudly pointing to our valorous young men upon fine horses, and to say: "Look at our cavalry!" and the great majority of the officers, not knowing themselves how to handle a sabre, can, of course, not impart its use to their men. Thus dash, valor, and horsemanship are absolutely thrown away as soon as the revolver has been emptied of its loads, and skilful swordsmen are closing upon them. It may be easy to say that such things will not occur, because we have to contend against enemies who do not even know how to ride, and the nature of our country forbids our ever employing large bodies in an actual charge in battle; but is it less necessary to be prepared against an inferior antagonist, or is it less requisite to be as skilful in all the requirements of a horseman upon the scout, the pursuit, and in grand *reconnaissances?*

The fact is, that a majority of the officers in higher commands in the cavalry service appear not to be impressed with the importance of their arm, and are either too indolent, or do not know how to instruct. Nor has our cavalry, in general, as yet been reduced to that degree of subordination and discipline, without which nothing important can be achieved. Otherwise, we ought to have a magnificent cavalry, after three years of warfare, as we have splendid infantry and artillery.

An important step has been taken in organizing our cavalry into divisions. Larger bodies of cavalry are thus brought together, and emulation is excited; the duties of each regiment, moreover, become less burdensome. If it is true—of which I am by no means persuaded—that we, on account of the nature of the ground, have to let our infantry and artillery alone sustain the shock of battle, and thus must give up the best part of a victory, we might just as well dismount the greater part of our cavalry force, and employ the rest in scouting and piqueting alone; because then even grand *reconnaissances* would become *dangerous.* With this organization into divisions, a step has been taken in the right direction, and we may, perhaps, soon hope to see the cavalry perform an efficient part in the shock of the great battle.

With these strictures it is not intended to say that our cavalry has done nothing. No man could be more sensible of its achievements under such leaders as Van Dorn, Stuart, Morgan, Hampton, the Lees, Wharton, Wheeler, Forrest, and others, than is the writer; but the *great* and *most magnificent* duty of cavalry, the *routing of a beaten enemy*, has not as yet been accomplished. No instances are necessary of what cavalry *can* do; history furnishes thousands of them; and if one would contend that the nature of the ground with us is such as to forbid anything like a charge or effective pursuit, he shows but ignorance as to the topography of European battle-fields.

6*

One fact should always be before the Confederate cavalry officer, namely: that he has under his command as fine riders and as splendid shots as the world produces— that he ought to perfect his men in discipline, the use of the sabre, and to precision in drill; and then I sincerely believe that he might outstrip the most renowned exploits of the arm.

And here I beg to say one word in favor of the "queen of arms"—the lance. One of my friends, a master of the sword and a skilful handler of the lance, has essayed to draw attention to the weapon and to form a body of lancers, and he has failed, as I understand. Furthermore, from high authority, I heard that some of our superior officers serving with the Army of the Potomac, in the beginning of the war, tried to introduce the weapon, and that they also failed, because the troops did not favor it. While we have this aversion for a formidable weapon, our enemies, it is said, are now actually organizing a strong body of lancers. This latter fact will, perhaps, more surely than anything else lead us to inquire whether it be not possible to arm the Confederate cavalry with lances.

If our men have no desire to employ either the sword or the lance, it is not because they dislike to use it, but because they feel that it is a harmless tool in their hands, since they are not taught how to use it. Did we leave the choice of the weapon to our infantry and artillery, I have no doubt that a good many infantry soldiers would prefer the light fowling-piece to the heavy musket, and the artillery-men would never bother themselves about heavy guns. I believe, on the contrary, that our men would like the lance, were its utility and efficiency once shown to them practically, as a man becomes attached to his sword as soon as he feels himself a master of it. It would, I believe, be a beneficial move if a good officer of cavalry could obtain authority to take five hundred young volunteers from the different cavalry regiments, and, *by a system of thorough drill and discipline*, make them a body of perfect lancers. Neither artillery nor squares have stood before them. Stationed on our coasts, especially, where the features of the ground favor their charging and pursuing, I believe them to be invincible. In a *mêlée* the lance is replaced into the receptacle near the left stirrup, and, by means of the straps, secured to the person of the lancer in such a manner that it does not interfere with the use of the sword, which then does its work. Then carbines and revolvers are useless, because they can not be reloaded.

Its manufacture is very easy. Every blacksmith can fashion a pike, and the trooper can himself fasten it to the staff. And the little flag, waving and fluttering in the breeze, alone has caused many a body of cavalry to turn upon the approach of lancers. An actual charge of lancers must be seen to appreciate what terrible effect this magnificent weapon has upon the steadiness of the horses and the *morale* of the men.

I believe it to be a matter worthy of inquiry:

1. Whether the tactics of cavalry should not be simplified as much as possible?

2. Whether the iron scabbard ought not to be abolished? since both the rattling it occasions, and the constant blunting of the sword's edge in a scabbard of metal, make it objectionable. Wooden scabbards might be substituted. A cavalrist might just as well carry a poker as a blunt sabre, for both are equally harmless. Another matter threatens to destroy the efficiency of the cavalry service entirely: the hardship imposed upon cavalry soldiers, and all mounted officers, to furnish their own horses, which daily is becoming more difficult, and will soon, save to a favored few, be impossible. There is no doubt that the government should furnish to all mounted men and officers their horses, as is the case in all well-regulated military establishments.

The last Congress was urged to provide for this emergency by the Honorable Secretary of War, in his very able report to the President, of November 26, 1863, but has failed to do anything.

It will be one of the most important duties of the coming Congress to save the cavalry service, which will soon become utterly inefficient unless something is done in the matter.

2. Marshal Murat.—Joachim Murat, ex-King of Naples, was born at a village in Pirigord in the year 1767, his father being an aubergiste, or country innkeeper. Prior to that, the father of Joachim had been a steward to one of the Talleyrands, and, through the influence of that distinguished family, young Murat was placed at the College of Chaors, and intended for the Church. His character, however, little fitted him for that profession, and he enlisted into a regiment of chasseurs. From this he was, within a brief period, dismissed for insubordination. He returned to his native village, and took charge of his father's horses until the breaking out of the Revolution, when he entered the constitutional guard of Louis XVI, from which he passed, as sub-lieutenant, into a regiment of chasseurs. During the Reign of Terror he professed himself an enthusiastic champion of equality, and rose to the rank of colonel; but these predilections did not prevent him from making himself useful to Bonaparte in 1795, and he was rewarded by being placed on the personal staff of the future emperor in the Italian campaign of '96. From that hour the fortunes of Murat closely followed those of his patron. The fiery valor which the handsome swordsman, as he was called, showed in a hundred fights, the splendid though somewhat fantastic costume in which he delighted to figure, and the love of daring achievement, which drew an air of ancient romance over all his actions, invested him, in the eyes of his admiring fellow-soldiers, with the renown of some paladin of old; and his enterprising talents in the field obtained for him the graver distinction, in the cool judgment of Napoleon himself, of "The best cavalry officer in Europe." He commanded that arm in the campaigns of Egypt, Italy, Austria, and Prussia; and in all —at Aboukir, Marengo, Austerlitz, Jena, Eylau, and Friedland—his services were brilliantly conspicuous. After the Egyptian campaign he obtained the hand of Caroline, youngest sister of Napoleon; and in 1806 was raised to the dignity of a sovereign prince, and recognized by the Continental powers as Grand-Duke of Berg and Cleves. In 1808 he commanded the French army in Napoleon's invasion of Spain, from which country he was recalled and sent to Naples to ascend the throne of that kingdom, vacated by the elevation of Joseph Bonaparte to the Spanish crown. In 1812 he accompanied Napoleon on the expedition to Russia, in the command of the cavalry of the grand army—the most numerous and splendid body of horse, perhaps, which the world has ever seen arrayed in the ages of civilized warfare. At the Battle of Borodino Murat performed prodigies of valor; at the same time, Napoleon (who was standing on a scaffold, which gave him a complete view of the *champ de bataille*) had his misgivings as to the over-impetuosity of his brother-in-law. That battle, one of the most ensanguined on record, opened the way to Moscow. But this city was a fatal possession to the French—driven out of it, as they were, by fire, to experience all the horrors of the Russian winter. Murat served the emperor again in the campaign of 1813; but, after the disastrous Battle of Leipzig, finally deserted him in his fallen fortunes, and became the ally of his foes. This unpardonable act of weakness and treachery was followed by another. Though he had saved his throne for a time, the hesitation of the Congress of Vienna to acknowledge his regality alarmed him; and when Napoleon reappeared in France, in 1815, he used his utmost endeavors to induce the Italians to arm for their national independence. But in this he most signally failed, and was compelled to fly. In a fit of desperation he again landed in arms on the coast of Calabria, with a handful of followers, and was captured and ignominiously shot by sentence of a Neapolitan court-martial—a base and most unwarrantable act, but worthy of its perpetrators. The people of Naples loved him much, for he was always mild and merciful as a sovereign. That he was one of the best cavalry officers of modern times there can not be a doubt; but it is equally certain that he was wholly unfit for a *chef d'armée*. He was a strange mixture—brave, vacillating, faithless, vain, wholly devoid of public principle, with a warm heart and kindly feelings.

3. General Nansouty was the general alluded to by Marshal Marmont. The French army was unfortunate, at the crisis of the Battle of Wagram, in losing the services of the gallant Bessières. General Nansouty permitted the Austrians to retire in the most perfect order. Napoleon, although this victory laid prostrate before him the Austrian empire, in a fit of ill humor exclaimed: "Was ever any-thing seen like it? We have gained neither any guns nor prisoners. The day will be without any decisive result!"

CHAPTER II.

ARTILLERY.

Importance—Simplest the best—In what perfection should consist—Necessity to restrain number of calibres—Character of siege artillery, and for defence of places—Inutility of 16-pounders—Field artillery—Conditions—6-pounders—Made the wars of the Empire—8-pounders, principal inconvenience—12-pounders, their object—Batteries of 24-pounders—Services upon battle-field—Hollow projectiles—Large mortars—Mortars Marmont *à la* Villantroy—Mountain artillery—Rockets *à la* Congrève—Wall-pieces—Dimensions—Weight—How determined—Theory of explosion—How range is diminished—Friction, expansion—Long pieces and heavily charged—Experience with 35-pounders—Maximum of range—What length—Inconvenience of length, giving greatest range—Length of 22, 28, 24-pounders—Inconveniences of instantaneous inflammation of powder—Particular phenomenon—Powder of General Rutti—Extraordinary force—In two schools destroying all pieces—Causes of this fact—Better the enemy of good—Weight of pieces—Force of recoil—Analogy—Experience of 1802 and 1802—Founded the system—Artillery of the Empire—Comparative weight of cannon and ball—Lightness of English pieces—Uniformity of construction of material—Importance—M. de Gribauval—His principles badly understood—Twenty-two kinds of wheels reduced to five—Unit of combat—Composition of battery—Its three elements—Importance of mobility—Its actual perfection—Work of M. Vallée—Artillery drivers—Made all campaigns of the republic—Artillery train, creation of author—System perfected under the Restoration—Artillery of new invention—General considerations—Congrève rockets—Views of author upon their employment—New combination of arms for battle—Two infantries—Proportion of arms—Infantry—Artillery—Its importance when cannon can not be employed—Mountain warfare—War on the plain—Effects of rockets in the first kind of warfare—Experience to follow—Necessity to be prepared for their employ—Revolution in the art of war—Paixhans' artillery—Character of heavy artillery—Swiftness compared with mass—System preferred—Distinction—Mobile artillery—Fixed artillery—Analysis of advantages of Paixhans' artillery—Resistance of air—Quantity of motion—Destructive action—Paixhans' cannon in the defence of places—Will change marine system—Ship of the line compared with frigate—Historical of Paixhans' guns—Influence of author upon their adoption.

Notes.—1. Remarks on Confederate artillery. 2. Armstrong, Whitworth, and Blakely guns. 3. Brass ordnance. 4. Deflection of Armstrong guns. 5. Projec- tiles for Armstrong guns. 6. Restraining the recoil of the Armstrong gun. 7. Shrapnel or spherical case. 8. Double-shotting. 9. Penetration. 10. Ricochet fire. 11. Salvo. 12. Spiking and unspiking guns. 13. Glass hand-grenades. 14. Grummet and junk wads. 15. Gun cotton or pyroxyle. 16. Carcass. 17. Lasso harness. 18. Slow-matches. 19. Ice. 20. Passage of rivers. 21. Pontoons. 22. Distance by sound. 23. Tactics of English artillery. 24. Present artillery tactics of the French. 25. Napoleon's organization for offensive warfare and plan of at- tack. 26. Table showing the *personnel* and *materiel* of a Prussian, Austrian, Rus- sian, French, and English light battery.

The third arm which has become indispensable in warfare is artillery. It is of capital importance; but its efficiency depends particularly upon its organization, and the principles upon which the latter is based.

I will endeavor to establish these principles and to develop their consequences. I shall first commence with the material, and will then pass to the means necessary to make the best employment of it.

The most simple artillery is the best. If one calibre could satisfy all requirements, and if the same carriage could serve for all different transports, the perfection of artillery would be attained.

But it is not thus. Artillery must produce many different effects; these effects being determined, it is necessary to find the calibres which can produce them, while limiting their number to strict necessity; because, so soon as one calibre can serve for the same object, there is one too many, and it becomes prejudicial, on account of the complications it leads to in regard to ammunition, changes, and repairs.

Artillery should be of three kinds: siege and fortress artillery, field artillery, and mountain artillery. In each of these divisions, and despite the difference which necessarily exists in the weight and dimen- sions of the several pieces, yet, as much as possible, the same calibres should be adopted, that the same ammunition may be used.

In sieges and the defence of places, pieces are necessary which kill the men, dismount the cannon of the enemy, and which are of great range. Experience has demonstrated that 12-pounders attain this object perfectly well.

In this species of warfare it is, besides, requisite to destroy the ram- parts, crumble them to pieces, and to open a practicable way to penetrate into the stronghold.   Here artillery is no longer a deadly weapon, but simply a tool, a machine—the battering-ram of the ancients; it has only become more powerful and expeditious.   To obtain this effect, 24-pounders are absolutely indispensable.   Sixteen-pounders, formerly used, have, therefore, become superfluous—insufficient in one case, and excessive in the other.

Field artillery should follow the troops in all their movements, and, promptly arriving upon a fixed point, should crush the enemy.   A light, easily transported material is, then, necessary, which being of the greatest mobility, need not stop before any obstacle the field may present. I believe 6-pounders, used all over Europe, and which I adopted when at the head of the French artillery, are sufficient for this purpose.   With this calibre all the wars of the Empire have been fought.   The 8-pounders have been again introduced.   There is no doubt that their superiority gives some advantages; but it is a great inconvenience to augment the weight of the ammunition by one-third, thus requiring more considerable means of transportation—means which are always wanting in war.

A second object in field manœuvres is to produce great effects with the aid of powerful reserves; to silence the fire of field-works upon which the enemy supports himself; to arm those which have been constructed; to open non-terraced walls, and to protect the passage of rivers.   For this, 12-pounders are required, but less heavy than those employed in sieges or in the defence of places.   Finally, every army should be accompanied by one or two batteries of short 24-pounders, to be fired with a less charge than one-third of the weight of the ball, and which, in a thousand cases, will render the greatest services on the day of battle.

The calibres, as just considered, should then be in accordance with the effects to be produced; and notwithstanding their great number, they can be reduced to three, by varying the dimension and the weight of pieces.

But this is not all.   Hollow projectiles, bombs, and howitzer shells are also used; it has been sought to adapt their calibre as much as possible to that of cannon, and no difficulty has been experienced.

The shell of a diameter of five inches, five lines, which has the same diameter as the 24-pounder, is used everywhere, with the advantage of being equally suited for cannon and howitzer.   A heavier calibre has been found useful for siege howitzers, and, in accordance with system, a diameter of eight inches has been given to them, which permits the employment of these shells in the 8-inch mortars, so useful in the attack and defence of places.

We come now to other mortars of superior calibre.   Here the greater the calibre, the greater the effect.   The expense and difficulty of the transportation of ammunition are the only arguments against their use. The mortars destined to receive a very heavy charge, cast upon a plate which supports them, and which were formerly called after me, as well as those called Villantroy mortars, are only applicable for coast defence, on account of their immense weight, and because their particular object is to obtain a very great range, which is useless in sieges and in the defence of places.

Later, I shall speak of a newly-invented artillery, which, while preserving the principle of unity of calibre, can be employed to produce different effects.

The calibres of which I have just spoken are, then, the only ones which siege and field artillery requires.

Finally, we come to artillery suitable for mountain warfare. Without entering into details, I will remark that it should be composed of pieces light enough to be carried on the back of mules ; heavier pieces, which are transported upon carriages, are more embarrassing than useful. Congrève rockets are also eminently suited for mountain warfare. I shall speak of this invention hereafter.

There exists yet another weapon which could be most advantageously employed; it is the wall-piece, but recently introduced, which is charged at the breech, and throws balls of several ounces weight with great accuracy, ranging equally with pieces of small calibre. These guns, distributed to the number of ten or twelve in each regiment, and placed, with their ammunition, upon a single carriage, would occasionally be of extreme utility.

After having spoken of the calibre of guns and the motives of their choice, it is proper to say a word as to the other dimensions of pieces, and their weight. Their determination is not arbitrary ; it is derived from positive circumstances, which directly influence a good service.

The length of a cannon depends upon the charge which is used. Experiments have not shown the precise limit for the greatest range; it has not been obtained in order to avoid other drawbacks, but it is pretty nearly determined. The gas which is formed by the ignition of the powder, and the explosion of which produces the force which pushes the ball, operates as a spring does when set free ; or, in proportion of its action upon the ball in movement, it augments the force which propels it, and consequently the range. This action is the result of ignition. If the ignition is not complete when the ball has left the cannon, there is a diminution of range ; if it is completed prematurely, and the ball has received the whole impulsion before having traversed the whole length of the cannon, there is also a diminution of range ; but then it is the friction which occasions it. The quantity of the powder must be such that this expansion of the gas which it produces when inflaming, accompanies the ball from the bottom of the breech to the mouth, neither more nor less ; thus, with long pieces stronger charges are necessary, and with shorter pieces weaker charges must be used.

In France, a uniform charge for cannon has been adopted, represented by one-third of the weight of the ball. With this charge, a series of experiments for the determination of the length which gives the greatest range has been made, and pieces of thirty-five calibres in length have been cast.

After having noted the range obtained, the chace of the gun has been diminished by sawing off the length of one calibre, and the range was found to be greater. The operation was renewed, and the result always the same, until reaching twenty-seven calibres; and, when passing this limit, to twenty-six calibres, the range decreased. It was then concluded that, with a piece of twenty-seven calibres in length and a charge of one-third of the weight of the ball, the maximum range was obtained.

But, with such a length, the manœuvring of pieces is difficult, and to remain within the mean of the limits for siege and fortress pieces, that of twenty-two has been adopted. For field-pieces, which require yet easier and more prompt manœuvring, this length has been reduced to eighteen calibres; in foreign countries it has been fixed at fourteen.

I do not speak of howitzers, an arm particularly designed for ricochet firing, established upon different principles, and required to fulfil different conditions.

I will now make a remark founded upon a well-established fact, the application of which is important, and which will be a matter of astonishment. Powder should inflame with rapidity, but not instantaneously; otherwise the inert force (*vis inertiæ*) occasions a shock of such violence as to destroy the gun itself. Its action must be successive. A particular fact has given me the means to examine this phenomenon.

General Rutti, an officer of great merit, and placed at the head of the Department of Powder and Saltpetre, had succeeded in manufacturing powder of extraordinary strength, and he believed to have obtained a very important object. Five hundred thousand pounds of this particular powder had already been made, and it was determined to preserve it as a precious article for times of war. Luckily, circumstances changed this destination. The new powder was ordered to be consumed for the exercises of the guard in 1828. In two schools all the cannon were burst and unfitted for service. These facts ascertained, I sought their cause, and no other explanation but that which I have given could be obtained. In this case the adage may well be repeated : "Better is the enemy of good."

As to the weight of pieces, it can be diminished very considerably and without inconvenience, with regard to resistance ; but the gun-carriages suffer by it, and are easily broken. The force of the recoil, operating upon too light a mass, produces a brisk shock, and destroys it. After the weight of the piece has reached a certain limit, whatever the excess of the weight of the metal, the gun-carriage should be strengthened to that extent.

This fact will be understood through the following example, which any one might see at any day: A juggler places upon his breast a stone of great weight, and braves the effect of a blow with a club, while, should this blow fall upon a smaller stone, he would be wounded.

In 1802 and 1803, when I was engaged in establishing the new system of artillery, which has served during the whole period of the empire, experiments which I ordered upon the weight of the metal demonstrated that that which satisfies equally the requirements of mobility and conservation is a weight of one hundred and twenty pounds per pound of the weight of the ball, well propelled by a charge of one-third of the weight of the ball; so that a 6-pounder gun, for instance, should weigh $6 \times 120 = 720$ pounds.

The English have attached great importance to the lightness of field artillery. They do not give, at least they did not give, thirty years ago, more than ninety pounds per pound of the weight of the ball; but they had likewise diminished the charge to one-fourth instead of one-third.

Another word upon the material. Gun and other carriages are necessary elements for artillery service; their use alters and destroys them, and replacements become constantly necessary. Hence the immense advantage of a perfectly uniform construction. M. de Gribauval, first inspector-general of artillery, author of the first regular system, has had the glory of establishing this uniformity. Thus the remains of a carriage constructed at Auxonne or at Toulouse can serve to repair a like carriage constructed at Strasbourg. But, left to the influence of the officers of the workmen, their pedantry led to useless divisions and subdivisions in the construction of material, thus occasioning, in a systematic manner, renewed great embarrassments, almost equivalent to the confusion from which the service had just escaped.

To give an idea, I will cite but one fact which has remained in my memory. There were, as well as I can recollect, twenty-two kinds of wheels in his system of artillery. In the system of 1803 I reduced them to ten. Nowadays four or five are only used; and I believe that no material has ever attained a like perfection. In the first war fifty pieces of cannon, well commanded, had more effect than one hundred, such as they were formerly.[*]

From the praise I have accorded to modern artillery, I only except the calibre of eight, which has been again introduced, and its exaggerated weight for field-pieces has again been fixed at one hundred and fifty pounds per pound of the ball.

---

[*] Marshal Valée, formerly central inspector of artillery under the Restoration, is the author of the fine system of artillery, in *personnel* and *materiel*, now adopted in France.—*Note by Author.*

But the best material in the world produces but an indifferent effect, if it is not placed in the hands of men who can properly manage it; and however remarkable the instruction of the corps of artillery has always been in France, many things were still wanting; its organization was very imperfect.

The most glaring faults have been successively remedied, and to-day the requirements of a well-established service appear to have been satisfied.

The unit of combat in artillery is the battery. It is composed of from six to eight pieces, always marching together, with their ammunition, and placed under the same command. It is in artillery what the battalion is in infantry—the squadron in cavalry. This body must, then, be homogeneous and compact; the elements which compose it must be organized in the same spirit, and be accustomed to act together.

There are three distinct elements: the material, or the guns, properly speaking, those who serve them, and those who conduct them. If these elements do not thoroughly accord, then artillery is imperfect.

The first merit of artillery, after the bravery of the cannoniers, is accuracy of fire and mobility. It is, then, easily seen of what importance is the management of the horses charged with drawing the cannon.

Formerly, everything was divided; the cannon remained in the arsenal or in park until the moment of an engagement; the horses belonged to a contractor, and the conductors were his servants, treated without consideration, having no prospects of fortune whatever, and called *drivers* of artillery.

With this monstrous organization the whole Republican campaigns were made.

Under the Consulate and during the Empire this service was raised, and the corps of artillery trains, with its non-commissioned officers and officers, was formed. Thus the prospect of advancement was held out, and the name *driver* was replaced by that of *soldier of the train.* My influence acted directly upon this organization—it was mostly my work; and, in order not to offend the rights of grades for the command, I took care to give to the officers of the train only very inferior grades in comparison with those of the corps of artillery.

Thus was prevented (which is indispensable), namely: all embarrassment and conflict in the relations of chief officers of batteries and those who conducted them. The latter, on account of their inferior instruction, should never have superior authority, and this difference in the grades placing them properly upon the list of military gradation, kept awake the spirit of duty and obedience. This organization lasted during the entire period of the Empire. At the end of the Restoration the Council of War, of which I was one of the vice-presidents under the Dauphin, changed the organization of the corps of artillery. It was divided into batteries, having their material, cannon, and horses conducted by cannoniers of the second class, who were at the same time instructed to manœuvre and serve the cannon, and which were called *cannoniers—conductors*. This organization has certainly attained perfection.

During the last few years two kinds of cannon have been invented, whose effect, according to my understanding, will be wonderful, if they are properly employed in the first war : the Congrève rockets for field service, and guns called Paixhans for the defence of coasts and fortresses. I firmly believe that the resistance of the latter will be increased. The conduct of war and the organization of armies will, likewise, experience a great modification. But these two objects merit particular development.

The part which artillery plays in warfare has acquired, with every day, more importance, not only by reason of its augmentation, but also on account of its great mobility, which permits its movements to be infinitely combined. However, there are limits to that mobility which gives the means of assembling upon a given point a great amount of artillery. The number of cannon which can be carried into war is equally limited, on account of the expenses and embarrassments an excess of material would produce—embarrassments of such extent upon marches, as to greatly surpass any advantages to be derived from their use in the moment of action. Experience has demonstrated that the *maximum* should be four pieces for one thousand men ; besides, this proportion will, after a few months of a campaign, be greatly exceeded, since the material is not subjected to the same causes of diminution as the infantry and cavalry, and the *personnel* of artillery being less numerous and, therefore, more easily kept at its proper complement.

But the Congrève rockets, which have, by degrees, arrived at great perfection, and which are now directed with sufficient accuracy, make an artillery which, through the development susceptible in their application, may become a principal arm.

Indeed, when the weapon only consists of projectiles ; when no machine is any longer necessary to throw them, and when no longer any surface is presented to the enemy upon which he can direct his shots ; when, finally, by means of very simple dispositions, an instantaneous fire can so be developed as to cover the whole front of a regiment with a shower of balls, representing the fire of a battery of one hundred pieces of cannon—then the means of destruction are such as to make any attack, such as the rules and principles of actual warfare have prescribed it, an utter impossibility.

I would conceive the employment of the Congrève rockets in the following manner : I would instruct five or six hundred men in each regiment in the service of this new arm. Two chariots would suffice to carry one hundred rocket-frames such as the Austrians have adopted, and at the word of command these one hundred rocket-frames, each served by three or four men, would develop such a fire as can hardly be conceived.

To a fire like this could you oppose masses, even troops in line of battle, upon several parallel lines? Assuredly not. But the gain of the battle consists in making the enemy recoil; the space which separates us from him must then be traversed by marching, and to do this with the least possible danger, that arm must be employed which can run over the distance with the greatest celerity. Hence cavalry suits best; and this cavalry even will be subjected to a new mode of manœuvring, in order to face the enemy's fire with the least chances of destruction. Therefore it must be scattered as skirmishers, ready, however, to unite at a given signal, to prepare for the shock which must follow the charge. Infantry then changes *rôles*; it becomes the auxiliary of the Congrève rockets, or rather these rockets become its weapon, and guns are only accessory.

In this new system infantry will have need of an entirely different instruction. It will be divided into two parties: the first charged with the service of the rockets; the second, in order to support the former, and to serve for its rallying point the moment it will come into immediate contact with the enemy. The proportion of the arms must then be changed; more cavalry than infantry will be needed; a cavalry which is drilled in a special manner, and an *infantry-artillery*, if I may express myself thus, whose employment will be limited to the service of the rockets, to sustain and protect them, to occupy entrenched posts, to defend fortresses, and to carry on mountain warfare.

But this new artillery assumes great importance in a thousand situations where cannon are of no use whatever. In mountains, a few pieces with great difficulty are now carried, which produce but little effect. With rockets we have a long range weapon, which can be everywhere established in great profusion, upon the crest of rocks as well as upon less elevated positions. In perfect plains every edifice is transformed into a fortress, and every village church roof becomes at will the platform for a formidable battery. In a word, this invention, such as it is, and as improvements will make it, lends itself to everything, is suited for all circumstances and combinations, and ought to have an immense ascendency over the destiny of the world.

Served by a special corps, and considered purely as artillery, the employment of the rockets would be necessarily rare, and would produce but little effect. An immense development is the only way to make them useful, powerful, and a matter of astonishment; they should, then, properly speaking, become the arm of the army.

The nature of things is but slowly considered. Routine acts a long time without employing itself with any possible modifications and ameliorations; therefore the power of Congrève rockets will not be appreciated but in a long period. But if, in the first war, a skilful and calculating general would consider the question in all its bearings and consequences, if he would quietly prepare his means to display them upon the battle-field, he will obtain such successes as defy all resistance until the enemy would employ the same means. At the moment of this grand trial the personal genius of the chief will have a great ascendency over the issue of the war.

However seemingly rational the result may be which I predict, experience alone will incontestably establish the merit of this new invention. The wise man will feel no absolute conviction until facts have realized his expectations, so many are the unforeseen circumstances modifying the deepest calculations and the most seducing probabilities.

Taking everything into consideration, appearances are such that a skilful and enlightened general should, in the first war, prepare to employ this new arm to astonish the enemy by its effects. If he uses it altogether, he will probably be master of the campaign; if his adversary has been as vigilant as himself he will, at least, guard himself from defeat. But his foresight should embrace all consequences of this new agent, considering its relations to the other arms—their proportions, manœuvres, and manner of being served.

After the successful employment of Congrève rockets during a campaign, it is evident that they will be adopted by all armies; then the equilibrium will be re-established, and none will have an exclusive advantage. But the art of war will be powerfully modified. More lively actions and their greater moral effect will make the battles shorter, and diminish the effusion of blood—since it is not the number of killed which gives victory, but that of those who can be frightened.

I repeat that Congrève rockets should create a revolution in the art of war, and they will be the first success and glory of him who, before any other, has developed their importance and the advantages to be derived from them.

I come now to the Paixhans guns.

Heavy artillery, to fulfil its purpose, should have great range, and the projectiles it carries a great quantity of motion. To obtain the latter, one of two things is necessary: either the velocity must be very great and the projectile less heavy, or the projectile must be very heavy and have less velocity—since the quantity of motion of a body is equal to its mass, multiplied by its velocity.

Until now, a less heavy ball with considerable velocity has been preferred, on account of the difficulty of transporting projectiles. But if this was right during sieges, when the means have to be transported in a short and fixed period, it was wrong under different circumstances,

7*

where sufficient time could be given to make transportation easy, whatever the weight of the material. For the defence of fortresses, the armament of coasts, and marine service, this artillery possesses immense advantages, which I will analyze in a succinct manner:

1. The resistance of the air to the motion of bodies being in proportion to the square of the velocity, it is much less with these projectiles, and hence both range and accuracy of fire are greater. Supposing a velocity of twelve hundred feet per second to be that of the ordinary ball, and four hundred feet that of the Paixhans ball, the resistance of the air will be as nine to one.

2. The quantity of motion of a twenty-four pound ball with a velocity of twelve hundred feet, will be represented by the number 28,808; while the Paixhans ball, of the 12-inch calibre, or of one hundred and forty pounds weight, with four hundred feet of velocity, will be expressed by 56,000—nearly double; that of a thirty-six pound ball, with the same velocity of twelve hundred feet, will be 43,000—consequently much more feeble.

3. The action of destruction being like the surfaces of the square of the diameters, the proportion will be one to four.

4. Finally, the thirty-six pound ball traverses the breastwork of an earthwork, or the sides of a vessel, or it buries itself. Wherever it may lodge, it causes no damage; and if it penetrates a plank, the hole is easily stopped up; but the Paixhans projectile produces different ravages. By its great diameter and the slowness of its movement, with an equal quantity of motion, the effect is contrary to that produced by great velocity; it demolishes a larger surface, makes an immense breach, and if a battery is struck, it must be reconstructed; if a vessel, it sinks, without any possibility of being saved.

The defence of a place, supported by such means, raises it almost to the moral strength of attack; and the employment of this arm upon the sea against vessels will cause the disappearance of squadrons, and especially that of large vessels. The superiority of a ship of the line against a vessel of inferior class has two causes: the ship carries artillery against which the thickness of timber of a frigate can offer no resistance; and the latter carries artillery of insufficient calibre to affect a ship of the line. Thus a frigate is unable to make the slightest attack upon a vessel of the line, since the fire of the former has only danger for the crew, and influences the manœuvres, while the fire of the ship of the line destroys, besides, the opposing vessel itself, and can, in a moment, send it to the bottom of the sea.

But now a small ship, either steam or sail, of but inconsiderable strength, can carry one or two pieces, a single ball of which suffices to destroy the largest vessels; ten small vessels, each armed with two heavy guns, can quickly surround a vessel and make an end of it. In this case, vessels costing as much as 1,500,000 francs, offer no guarantee of security or exploits. The Paixhans artillery is, then, the destruction of navies as now constituted.

During the Restoration, Lieutenant-Colonel Paixhans, an officer of great distinction, conceived the idea of proposing this artillery. Louis XVIII nominated a commission of generals and admirals to examine it, over which I presided. The exposition of this system struck me by its novelty and just conclusions, and I became its declared partisan.

Experiments were, however, necessary to determine the range, greatest accuracy of fire, and manner of manœuvring this ordnance most easily. They were made at Brest, and succeeded perfectly, surpassing the expectations of the author. From thence date those changes in artillery which have immensely modified war upon the seas, in rendering large vessels superfluous; the defence of coasts has become easier and surer; and on this account, it appears to me, the defence of fortresses will be much prolonged. But the adoption of this new arm should not dispense with the employment of hollow projectiles, fired from 36 and 24-pounders—since their effects, though less powerful, are still formidable against an enemy, and favorable for defence.

NOTES.

1. Remarks on Confederate Artillery.—Hereafter, artillery officers of our service will have to give important testimony as to the determination of questions which are now being inquired into amid the shock of battle.

The increased range of artillery, still on the ascendency, and, in the case of heavy guns, apparently destined to assume a fabulous extent, and their relative effects upon armies and fortifications of all kinds, with the artillery officer; the best mode of resisting the effects of modern artillery by means of earth, stone, and iron, and those of iron-clad navies, with the engineer officer; and the influence of the improved small-arms upon men, courage, and tactics, with the officers of all arms—are still points susceptible of controversy; and our officers will be peculiarly apt to give opinions thereon worthy of consideration, since whatever science and military skill can bring to bear on the issue of the struggle is now being practically tested.

And, in view of these important questions at issue, it is gratifying to know that our officers of artillery appear to be fitted in a high degree for that duty. The artillery service of the Confederate States, more than any, has won laurels in this war. No one can witness the performances of our corps of artillery without being impressed most favorably. And because artillery officers feel at once that they must study and attend to their duties, or else their weapon, instead of achieving its high mission of breaking the confidence of the advancing enemy, becomes an encumbrance, they have earned the first place among the arms of the Confederate States; and any unbiassed critic, who has had the opportunity of seeing the achievements of European artillerists, must say that those of our army are not surpassed by any in point of mobility, accuracy of fire, and bravery of the men.

A different organization of our artillery into regiments, brigades, and divisions, so as to give a more rapid promotion to the corps, appears to be just and necessary. Now, a captain of artillery has the prospect before him of being a captain as long as the war lasts, except he be fortunate enough to be selected by his commanding general as the chief of artillery, with the rank of major or lieutenant-colonel.

It will be perceived that Marshal Marmont's chapter on artillery does not embrace that which would be most acceptable to artillery officers of the present day, namely: a treatise on rifled guns and siege pieces of all kinds. It is hoped that some of our artillery officers will think it worthy of bringing this chapter to the present day.

2. Armstrong, Whitworth, and Blakely Guns.—From the results of a very extensive course of experiments made at Woolwich, in the latter part of the eighteenth century, Dr. Hutton was enabled, with a view to the increasing of the velocities and ranges of projectiles, to recommend their being cast of a shape offering less resistance to the atmosphere than that of the spherical balls then in universal use—namely, shot of a long form—a hint adopted and improved upon by the modern artillerist inventor. The above celebrated mathematician also proved by these experiments that there is no sensible difference caused in the velocity and range of projectiles by varying the weight of the gun, nor by the use of wads, nor by different degrees of ramming; that *velocity*, with equal charges, always increases as the gun is longer, though the increase in velocity is but very small in proportion to the increase of the length of the bore; and the *range* increases in a much lower ratio than the velocity, the gun and elevation being the same. Hence it is evident that we gain extremely little in the range by a great increase in the length of the piece.

Our time is emphatically an age of progress in art and invention. The French now boast that they possess a muzzle-loading 8-pounder rifled cannon, for mountain service, weighing only two hundred pounds, with which shot can be projected five and a half miles; and the following tabular record of Mr. Whitworth's practice with his 3, 12, and 80-pounders, at different degrees of elevation, shows marvellous results both as regards range and accuracy:

Summary of experiments with Whitworth's rifled, breech-loading cannon, at South-port, showing the mean range and deviation of all the shots fired at each experiment.

Date.	Calibre of gun.	Elevation.	A. Number of shots fired.	B. Range.	C. Longitudinal deviation	D. Lateral deviation.
Feb'ry.		Degrees.		Yards.	Yards.	Yards.
22	3-pounder	3	10	1,579	12	.52
15	"	10	5	4,174	27	1.17
16	"	10	5	4,190	87	5.50
23	"	10	10	3,842	48	3.23
15	"	20	4	6,793	58	4.83
16	"	20	4	6,960	69	8.58
22	"	20	5	6,647	109	7.40
22	"	20	4	6,421	94	4.25
23	"	20	11	6,663	33	3.83
15	"	35	4	9,015	96	10.92
16	"	35	5	9,580	81	19.33
22	12-pounder	2	5	1,247	24	.85
16	"	5	5	2,324	11	1.57
22	"	5	10	2,336	16	1.08
23	"	5	10	2,219	22	2.09
21	"	7	4	3,049	17	.50
21	"	7	4	3,098	9	.54
16	"	10	5	4,027	50	3.31
23	"	10	10	3,774	37	3.10
15	80-pounder	5	2	2,575	36	2.33
...	"	5	2	2,574	30	1.66
23	"	7	4	3,493	8	.58
16	"	10	2	4,700	30	.50
22	"	10	4	4,409	50	5.17

REMARKS.—Column A shows the number of shots fired at each experiment; B shows their average range in yards; C their average longitudinal, and D their average lateral, deviation from a central point, according to the system adopted at Hythe.

All these experiments, be it remembered, were conducted from first to last by amateur gunners, without accident or delay of any kind, and during the prevalence of wind and weather both about the most unfavorable that could well be imagined for obtaining good average results of range alone.

Since the great success of the Whitworth gun was first published in the *Times,* that gentleman has continued to demonstrate the scientific principles on which his ordnance are constructed, by eliciting from fresh experiments a regular pro - gressive increase of range and accuracy. Mr. Sidney Herbert has, therefore, stated in the House, that as the Whitworth gun had exceeded the Armstrong in range, and very nearly, as far as could be judged from the rough experiments, equalled it in accuracy, the government were prepared to take the usual steps to give both a competitive trial at Shoeburyness.

A clever writer, in the *Army and Navy Gazette,* discusses the merits of the Armstrong gun thus:

" So far as accuracy is concerned, Armstrong's gun shoots to perfection. At three thousand yards, with a field-piece, you may go through a six-foot target to a certainty. If you miss, blame the gunner, not the gun. There is nothing more to be desired in range; for over an undulatory and unknown country there is not one man out of ten whose sight and judgment at that distance can be depended on.

"Nevertheless, there are serious blots in this celebrated cannon.   First and fore-most, it can not be used without water, and plenty of water too.   The Minister of War mentioned in the House that this difficulty has been overcome.   With all re-spect, I beg to say he is in error.   The greasing wad has improved matters, and that is all.

"In consequence, sponging is excessive, and takes long to accomplish—far longer than a field-gun can afford in action.   It must be thoroughly well done with water, or it is of no use at all, from the difficulty of inserting the bristles of the sponge in the grooving where the fouling lodges.   Can you always depend on water in the field of battle?   I think not.

"But what I think will be a fatal objection to these guns in action, is the expos-ure of the great breech-screw when loading.   The male and female threads of this screw fit with beautiful precision ; when untwisted in loading, the male threads are exposed to all the dust and grit which horses, men, shot, and other contingencies in action, will surely cause.   If it hitches in screwing up again, no power on earth will move it, and the gun must be abandoned or dragged out of action.   I saw a breech-screw once so misbehave without any apparent provocation, and when com-fortably shedded at Shoebury all the artificers failed to stir it; so it was sent to Newcastle to be expanded by heat; but, alas! we can not carry Newcastles on a campaign.

"Compared with old field-guns the whole affair is complicated, and will require unremitting attention from officers and men.

"This applies to field-pieces only; for heavy guns on board ship the disadvantages vanish; there dust and grit can be reduced to a minimum, and water is ever plen-tiful.   This is the field for Sir William.   I long to see a frigate fitted with his heavy guns.   Woe betide her adversary in the fearful day of battle."

It is alleged, by the supporters of the Armstrong gun, that Mr. Whitworth has only obtained greater range by reducing the diameter of his projectile, and, of course, therefore, the bore of the cannon itself; quite forgetting that, as long as that gentleman can prove that a great improvement is brought about by the adop-tion of certain principles, the public and the military authorities will care very little whether the principles themselves are new or old.   By Mr. Whitworth's plan of reducing the diameter of the shot, and, therefore, the bore of the gun, he con-tends that not only are the range and accuracy increased, but the gun itself can be constructed of the same relative strength of metal, though nearly two-thirds lighter than the ordinary brass guns.   The value of this reduction in weight, by allowing fewer horses and fewer men to manœuvre heavier guns at greater speed, must be apparent to any one, more especially to those who have seen what a very little way the largest transports in the service go toward transporting two or three ordinary field batteries, with their present complement of twenty-one wagons and carriages, two hundred and fifty horses, and some two hundred and fifty men.

The celebrated 3-pounder gun of Whitworth, with carriage and limber complete, could be brought into action, and manœuvred and served with the utmost rapidity, by two horses and two men only.   In this respect, however, the Whitworth gun has no advantage over that of Armstrong; on the contrary, as far as we have yet seen, the Armstrong large guns are much lighter.

It has been stated that the Armstrong 70-pounders can be worked by five men as easily as the old smooth-bore 68-pounder by seventeen; and the precision of fire attainable by it is alleged, by competent authority, to be fifty times greater than that obtained with the ordinary service gun, throwing solid shot of equal weight, each at a distance of one thousand yards ; indeed, it is said to be six times more accurate at three thousand than the ordinary smooth-bore gun at one thousand yards.

With reference to range, however, Sir William Armstrong himself states that, beyond a certain distance, range, for general purposes, has no practical value, and that as for artillerymen firing in the field at objects five miles distant, without any clue to guide them but their eye, they might as well fire at the moon. It is not only a question of which shot goes furthest, but what the shot effects when it does reach the mark. The formation of his gun, he states, has not been his chief or only object—which, in fact, has been as much directed to inventing the most destructive projectile.

To secure this all-important object he has been compelled to give up, to a certain extent, the attainment of an immense range, and increase the diameter of his gun in order to enable it to carry the Armstrong shell, which, for terrible destructiveness, deserves to be almost more celebrated than the gun itself. Thus he states that, as yet, no fair comparison can be drawn between the results he has achieved while trying only for destructive effect, and the results obtained by a gun which was merely fired for range.

The real test as to their merits, both he and Mr. Whitworth very justly maintain, can only be got by putting the two guns side by side, and trying them under similar conditions for range, accuracy, and, above all, for destructive effect.

The introduction into the service of either the Armstrong or the Whitworth effective and, comparatively, light guns, will probably prove a check to experiments for the construction of huge pieces of ordnance, such as Mallet's two thirty-six inch mortars, which, it is said, were made at a cost of £40,000, and one of them was disabled by a charge of forty pounds of powder, although the full charge should have been four hundred pounds. These would seem, however, to be mere pocket-pieces compared with one, the proper material for the construction of which was alluded to by Captain Blakely, R.A., in a lecture delivered at the Royal United Service Institution. The monster mortar contemplated by this gallant officer was to be of sixty inches calibre, to throw seven tons ten miles!

Sir W. Armstrong claims to have constructed his gun on certain fundamental principles; these Mr. Whitworth disregards, and forms his gun as unlike Sir William's in principle as two guns can well be.

The Whitworth gun, as distinguished from the Armstrong, is bored from one solid cylinder of homogeneous iron. There is no rifling, as is generally understood by the term, in the bore, which is a plain hexagon, making one complete turn, which varies with the diameter of the gun. Thus there is one turn in about eight feet in the largest guns (from 50 to 120-pounders), one complete turn in five feet in the medium-sized ordnance (12 to 32-pounders), and one complete turn in three feet four inches in the small guns, or from 3 to 12-pounders. All the guns above 18-pounders are hooped round with rings of iron forced on by hydraulic pressure—an additional strength which is apparently not required, and which, in weight, gives the Armstrong guns, of the same calibres, a most important advantage. The breech-loading arrangement is a hinge at the end of the gun supporting a hoop of iron, in which is the breech or cap, which screws on to the end of the piece. The shot is of cast-iron, and in form precisely like a ninepin, with its thickest part at the middle pared off, to fit with mechanical precision the hexagonal sides of the bore. Thus the projectile has a bearing surface on the whole of the barrel, and runs freely in or out of the gun, so that in case of an enemy's shot striking the breech and jamming the screw, or other injury to it, the gun could be used as a muzzle-loader with the same facility as an ordinary smooth-bore field-piece. We need scarcely say that this is not the case with the Armstrong; anything happening to the arrangement of the breech at once rendering the gun useless, till another breech is fitted on at the factories at Elswick or Woolwich.

With the Whitworth gun there is no chamber for the reception of shot and powder—an advantage of the utmost importance. The Armstrong chamber adds to the length of the gun, without being rifled, or assisting in impelling the shot in any way. With the Whitworth, the gun is rifled throughout its entire length from end to end, and every inch is used to aid the flight and give rotation to the projectile. From the chamber in the Armstrong being of a certain size, it follows that only shot of a certain length can be used. In the Whitworth, on the contrary, it is contended that shots of any length, or a charge of powder of any strength, can be used indifferently. Thus the 3, 12 and 80-pounders are, in fact, only guns of the calibre we mention, as long as they are required to throw a distance of five or five and a half miles. Reduce this enormous range to the distance at which long range guns are generally used—say three thousand yards— and the length of the projectiles of these ordnance may be more than doubled: the 3-pounder used for nine pound shot, the 12-pounder for thirty-two pounds, and the 80-pounder for a shot of even two hundred pounds. In naval warfare, great weight must be attached to these advantages. 12-pounder boat guns could be used as 12-pounders or 36-pounders, according to the distance at which they chose to engage, while ships could double-shot or even treble-shot their broadside guns as they closed with an enemy. The only limit, in fact, to the number of shots with which the Whitworth can be loaded when engaged at close quarters, is the limit to the strength of the powder to eject them. Thus, in the course of the experiments tried to ascertain this fact, it was found that the 3-pounder got rid of ten shots' placed one over another, at one discharge, but failed to eject eleven, when all the powder in the charge burnt out like a squib through the touchhole, leaving the shots in the gun.

Captain Blakely has lately advanced a claim to a portion of the world-wide fame which Sir W. Armstrong and Mr. Whitworth have achieved for themselves by the invention of their wonderful ordnance. He has constructed a new cannon eight and a half feet long, and weighing only forty-eight hundred weight, which projects shells of fifty-eight pounds weight to a distance of upwards of a mile and a half, with only five degrees of elevation—beating, it is alleged, Mr. Whitworth's 80-pounder solid shot gun, weighing eighty-four hundred weight, by one hundred yards, at the same elevation. It has also been asserted that as Captain Blakely's gun is six inches in bore, and Mr. Whitworth's only four and a half, the initiatory velocity of the shells from the former must be vastly superior, and their advantages at shorter ranges still greater than at the distance chosen for experiment.

We subjoin, as a further illustration, a debate in the House of Lords on the controversy between the Armstrong and Whitworth guns, on February 9, 1864, which will prove of interest to artillery officers:

"The Earl of Hardwicke wished to know whether Her Majesty's ships were supplied with any guns or projectiles that could penetrate the iron plates of a ship's side four inches and a half thick. This question had a most important bearing upon the warlike power of this country. By modern improvements the range of musketry had been so extended that field artillery was commanded by it. Inventors then set about devising means to meet this altered state of musketry range, and one invention was produced which excited the wonder of artillerists by length of range and precision of aim. He was a member of the Cabinet to whom the invention was made known, and after the most careful consideration of the experiments made and the results obtained, the government entered into negotiations with Sir W. Armstrong to superintend the manufacture of guns on his principle for the public service. The government then in power was careful to limit the manufacture to field guns, and he believed those guns were now regarded as very valuable weapons. By the peculiar mode of rifling adopted by Sir W. Armstrong, and the exact fitting of the projectile to the bore of the gun, breech-loading was necessary, and in that shape the Armstrong gun was undoubtedly an excellent weapon.

"Soon after a rival appeared to the gun in the form of iron plating for ships. Experiments were made, the result of which was to prove that the old 68-pounder gun was the most destructive weapon. Immediately afterward appeared another invention of cannon and projectiles upon a wholly different principle. He did not think the government had done justice to themselves nor to the inventor by the course they had pursued, although it was not unnatural when another inventor had been placed at the head of the government manufactory of cannon, and had expended about two and a half millions of money in producing weapons upon his principle. The Armstrong guns were found to be ineffective against iron plating, but nevertheless they formed a portion of the armament of our ships-of-war, together with 68-pounder guns of the old pattern.

"It appeared that Mr. Whitworth had not been permitted to carry out his experiments exactly in the manner he wished, although he was well known as a man who had devoted much attention to the subject of the manufacture of guns, and in 1857 had taken out a patent to secure his invention. At last Mr. Whitworth was permitted to make some experiments, and the House would understand the reason for the question he was about to put when he stated that Mr. Whitworth had never failed to penetrate the iron plating to which his guns were opposed, thus rendering our iron-plated ships no better than the old wooden ships. That fact should induce the government to allow Mr. Whitworth to show all that he could do; and if that gentleman succeeded in all that he undertook to do the result would be to relieve this country from a great expenditure, and to give us again a fleet of ships which could fight and float in the severest weather.

" In May, 1860, an experiment took place, of which he believed the noble duke was a witness. One of Mr. Whitworth's guns was fired against the Trusty, a ship plated with four-inch iron, originally built for harbor defence. The result of the experiment was that every shot from the Whitworth gun passed clean through the iron plating. He then brought a heavier gun, and the result was the same. Since then there had been important experiments at Shoeburyness against targets made exactly to represent ships' sides. Mr. Whitworth brought a 70-pounder gun and also a 12-pounder, and the effect, he understood, was marvellous. At every discharge his shot went clean through the target, and there could be no hesitation as to the result of his gun on a ship's side.

"He had spoken of solid projectiles; he now turned to shell. In 1862 Sir William Armstrong stated to a scientific society at Sheffield: 'It may certainly be said that the shells are of no avail against iron-plated ships, and that neither a 68-pounder nor a 110-pounder gun with solid, round, or conical shot is effective against them.' In May, 1860, the experiments against the Trusty took place, and he might conclude, from what Sir William Armstrong said, that there was no shell firing then. But Mr. Whitworth's shell went right through the target as if it had been paper, making frightful havoc of the interior lining of the ship. This distinguished manufacturer had always done what he had engaged to do, and if he had failed in any instance it was because he had been working with old cast-iron guns rifled by himself.

8

"The armament of the sea service Armstrong gun might not be useful on board ship, and would not bear the charge required for heavy artillery. The machinery was so delicate. and the fitting of the vent-piece so nicely adjusted to prevent the escape of gas, that a gun with anything like a quick discharge became useless in two or three rounds. Under these circumstances he had thought it his duty to put this question to the noble lord. It was time that the talents of such inventors as Mr. Whitworth should be made available by the country, and if his gun and projectile were such as he believed them to be, the more rapidly they were brought into the use of the public the better. The question was, whether Her Majesty's ships were supplied with a gun and projectile which at once would penetrate a ship's side plated with iron four and a half inches thick?"

"The Duke of Somerset, who was heard with great difficulty, said, as his noble friend's question immediately related to the navy, it might be more convenient that he should reply to it, having from the first been acquainted with and taken part in these experiments. When the present government came into office, in 1859, they found a record in the department strongly approving and praising the gun of Sir William Armstrong, and one of his guns had been ordered to be made. He was very anxious, as the public also was, that the navy should be supplied with a rifled gun ; he therefore communicated with the late Lord Herbert on the subject. A 70-pounder gun was sent to them in November. It was sent to sea under charge of Sir W. Wiseman, who was to report upon it. In point of accuracy, and every other quality required, it was pronounced excellent for all purposes. Accordingly, a certain number of those guns, which were to be 100-pounders, were ordered. In the meantime it was true, as the noble earl had stated, that the question of armor-plates arose; and it was found when tried at Shoeburyness that the gun, of which the accuracy and power against wooden ships was tremendous, had not sufficient power against iron plates. Neither the 68-pounder nor the 110-pounder could penetrate the iron plates. If anything, the 68-pounder struck the heavier blow.

"He then saw Mr. Whitworth, who said to him that he could produce a gun and projectile that should penetrate the iron plates. The projectile, he said, must be of a very peculiar manufacture, but it could be done. He accordingly communicated with Lord Herbert, who said he was most anxious to try the experiment. They went down the river to the Nore, and had the Trusty anchored two hundred yards off. The gun was fired, and undoubtedly the bolt went right through into the vessel, and had they gone on with two or three shots more they would have sunk the vessel. He proposed to Lord Herbert that they should buy that gun; they paid Mr. Whitworth a large sum for it, and they continued to try some other experiments. The misfortune was, that in some of their trials the gun was found to have a flaw, and eventually it burst. Still, he was in hopes that they should have some more guns supplied. They frequently tried to obtain other guns from Mr. Whitworth. He could assure the noble earl, so far from favoring any one manufacturer—Sir W. Armstrong or any one else—they were only anxious to get a gun that would answer for the navy.

"Rather more than a year ago he had further communication with Mr. Whitworth, who said the difficulty was as to the material, to get coiled homogeneous metal; for the gun produced was made at the royal factory at Woolwich. He repeated to him that they were most anxious to make a trial of his gun against that of Sir William Armstrong. From that day to this they had never got the gun.

"Mr. Whitworth's inventions were, he admitted, very clever, and he certainly had no favoritism for one inventor over another. All he desired was to get a good gun for the service. Whatever was the quantity of powder they put into the gun, if they fired with cast-iron shot, the effect was very trifling. Indeed, they might almost as well fire mud at the target, unless the projectile was of a very hard substance. No sooner, however, had they obtained a hard projectile than not only Mr. Whitworth's, but Sir W. Armstrong's, gun could fire a shot that would penetrate an iron plate. The iron plate committee, which had watched these experiments, had not reported that the flat end was of the least importance to the projectile. Yet they had been left in great difficulty as regarded the navy, because, although they had 130 and 150-pounders which would send a shot through an iron plate, they had not got a broadside gun that would answer their purpose.

"About a year ago he communicated with his noble friend at the head of the War department, and with his concurrence sent for Sir W. Armstrong, and told him that while they were going on with their experiments and trying various schemes for rifling, which might occupy their attention for months, and it might be for years, the Admiralty would never get a gun; that they really wanted a plain gun in the meantime for the use of the navy, which they might charge with from twenty-five to thirty pounds of powder. Sir W. Armstrong said that that could easily be furnished; and while such a gun was making he also suggested that the War department or Sir W. Armstrong should make a gun of from six to six and a quarter tons weight for the navy, for practical convenience required some limit to be put upon the weight. Accordingly that gun was produced in September or October last, and the results were very satisfactory. The practice made at from one thousand to twelve hundred yards was accurate. At twenty-six hundred yards the practice was still good, although inferior to that made by the rifled 110-pounder. The concussion between the decks did not appear to be objectionable; they found no difficulty in training the gun, and when tried in a gunboat it was as easily controlled as the 68-pounder. The gun was considered superior for service against iron-plated ships to any gun they had.

"The smooth-bore 110-pounder, with a charge of twenty-five pounds of powder, penetrated through and through a five and a half inch iron plate with round shot. That showed that, after all the talk about punch-headed shot, what they wanted was a good, hard, solid shot. The result, as far as it went, was perfectly satisfactory in regard to the broadside gun. A quantity of these guns had been made while the experiments going on were in progress, and some of them would be delivered in a very few weeks. As there was no complexity in their construction, no breech-loading or rifling, they could be made very fast. At the same time that was not all that they wanted. They required a rifled gun of about the same weight. When they got that they would see whether they lost any advantage by the rifled gun. One advantage they must lose was that they must have a lower calibre, and, instead of having a 9-inch bore, they would have one of about seven inches. They had, therefore, got a rifled gun of the same weight, and were only waiting for projectiles to be made for it. The greater nicety in the projectiles used for the rifled gun caused some delay; but he hoped in a few weeks to be able to give the House more precise details as to these experiments.

"The noble lord would see that although the guns with which the navy was provided were no more than the 68-pounder and the 110-pounder of Sir W. Armstrong, yet that they were in the way of making a gun which would answer broadside purposes, and also of getting a gun with rifling. The truth was that the whole question of the manufacture of iron was now in a state of transition. Every day they would see new experiments with guns and projectiles, and for this reason they wanted a projectile that would go through iron plates at a reasonable cost.   There were many qualities of iron, but they found that only the very best steel would pierce a plate satisfactorily.   There was little doubt that in a short time they would be able to send spherical projectiles through iron-plated ships.

"But when the noble lord said that wooden ships would therefore be as good as iron ones, he could not agree with him.   He was afraid that, whatever they might do, they must still keep to iron-plated vessels, because, although shots might pass through them, yet the inevitable destruction of wooden ships by shells would be such that warfare by such ships against iron-clads would be out of the question. Therefore, while they were increasing the force of their guns, they were also increasing the strength of their ship's sides.   The last specimen of a ship's side which they had was very considerably stronger than those they had before; and he hoped that the vessels so constructed would be able to go to sea and keep the sea in all weathers.

"He thought, then, that they had made all the progress in their power.   He was sure that no pains had been spared to do so.   The subject was naturally an interesting one, and no one could take it up without desiring to see the progress made in it.   Many highly intelligent minds were engaged upon it, and were continually sending in new inventions; and although the government were so overrun with new projects that if they attended to them all endless delay would ensue, they, nevertheless, desired not to overlook any valuable practical improvement.

"The Earl of Hardwicke did not think that what had fallen from the noble duke at all contradicted what he had stated.   It really appeared that Mr. Whitworth had long since succeeded in doing all that the noble duke was about to do with his gun.   In Mr. Whitworth they had a man who could produce a weapon and a projectile to penetrate a four and a half inch plate, and though they saw the effect of his shot and shell, the government declined to employ him.   The noble duke had himself turned inventor, and found a smooth-bore gun which sent a spherical shot through an iron plate.   But Mr. Whitworth's projectile was one of a most wonderful description, and with a raking broadside against an iron-cased ship it would be almost as effective as against a wooden one.

"He would earnestly caution the government against putting into the broadside of a ship a gun weighing six tons.   Their lordships would observe that his question had brought out an important fact, viz: that at the present moment, notwithstanding an enormous expenditure continued during ten years, the royal navy was not provided with a single gun capable of penetrating an iron-plated ship.

"Earl De Grey and Ripon thought the noble earl had somewhat misrepresented the relations between the government and Mr. Whitworth.   At an early period, as their lordships had been informed, the noble duke at the head of the Admiralty put himself in communication with Mr. Whitworth, and the gun which had been so successful against the Trusty was purchased by the government.   Some very important experiments were made with it in the autumn of 1862.   The most remarkable one took place at Shoeburyness, when, in September, Mr. Whitworth threw his shells through the Warrior target.   Almost immediately after that experiment, the communication between the government and Mr. Whitworth still continuing, it was proposed by the noble duke at the head of the Admiralty and the late Sir George Lewis that a fresh committee, composed of persons who had not dealt with the question previously, and against whom Mr. Whitworth could urge no objection on the score of prejudice, should be appointed to consider the merits of the Armstrong and Whitworth guns.

" Concurrent trials between the two guns had been proposed to Mr. Whitworth at different periods, but he had invariably objected either to the programme which had been proposed, or to the committee to which the inquiry was to be entrusted. His real objection was that he did not like the committee, because he thought some of its members were persons who had approved the Armstrong gun originally, and were committed either by that approval or by subsequent inquiries in favor of the Armstrong gun. In order to meet that objection it was determined, as he had already said, to appoint a fresh committee. A personal friend of Mr. Whitworth was added to the committee; another person was appointed to represent Sir William Armstrong; the instructions to the committee, after careful consideration, were shown both to Sir William Armstrong and to Mr. Whitworth, and at length the committee was formally constituted on January 1, 1863. It commenced by taking evidence. That employed it for about three months; but in the meantime it ordered a certain number of guns from each competitor. The guns so ordered were 12-pounders and 70-pounders. In a short time the 12-pounders were delivered, but Mr. Whitworth's 70-pounders had not been sent in yet, and from the time when it closed its evidence the committee had done nothing except repeatedly calling upon Mr. Whitworth to produce his 70-pounders. That was the reason why the inquiry had been stopped. Mr. Whitworth himself accounted for the delay by alleging the difficulties he experienced in getting the steel which he required. He could assure their lordships that there was not the slightest disinclination on the part of the government to inquire fairly into the inventions of Mr. Whitworth. The inventions of so eminent a man of science were mostly worthy of examination; but the government could not proceed hastily or rashly in such a matter, and all they wished was that Mr. Whitworth would produce his gun in order that it might be fairly tried."

3. Brass Ordnance.—The advantage of brass ordnance over iron, hitherto, has been that they were lighter than it was supposed—or known, until the invention of the Armstrong and Whitworth guns—iron ones could be cast of the same calibre, without risking their bursting. But it has long been known that the former are sooner rendered unserviceable by repeated firing than the latter.

From ten to twelve discharges in an hour are considered the greatest number that ought to be made from a brass gun.

In 1811, for the Siege of Badajos, the English army procured, from a neighboring Portuguese fortress, a battering train of brass ordnance; but, from rapid and continuous firing, the guns very generally failed—not by bursting, but by bending— the bores no longer remained straight. The failure was due to the property of the metal, which will not stand the amount of heat generated by continued explosions from large charges.

The great lightness, affording facility of transport, of the newly-invented Armstrong and Whitworth guns—their tenacity of material, to withstand a large amount of firing—precision of fire, extent of range, with reduced charges, and breech-loading, both contributing to lessen the heating of the pieces in action—are qualities which altogether must eventually, and indeed ere long, cause the entire displacement of brass guns.

4. Deflection of Armstrong Guns.—The Armstrong guns always throw to the right, increasing with the range; this is termed a constant deflection, and must be allowed for. The allowance is made by placing the rear sight so as to be correct at two thousand yards; at distances below two thousand yards, the error will be so small that it may be disregarded.

8*

> From 2,000 to 2,500 yards, give 3′ deflection left.
> From 2,500 to 3,000 yards, give 5′ deflection left.
> From 3,000 to 3,500 yards, give 9′ deflection left.
> From 3,500 to 4,000 yards, give 15′ deflection left.

To these must be added the allowances for wind, and this increases as the squares of the times of flight.

A practical rule is, that each minute of deflection on the sight gives a difference of an inch in every hundred yards of range.

5. Projectiles for Armstrong Guns.—One species only of projectile is used for field guns; this serves indifferently as shot, shell, shrapnel, or case.

The projectile is cylindrical, with a conical point; is exteriorly coated with lead, and contains forty-nine cylindro segments of cast-iron; in the centre a space is left to receive the bursting charge and concussion fuse. A female screw in the head of the shell receives the time-fuse or plug.

Molten iron can not be used in the Armstrong projectile, as it would melt the lead coating which takes the rifling.

6. Restraining the Recoil of the Armstrong Gun.—In a very recent experiment made before the select committee of Woolwich Arsenal, with solid india-rubber buffers—invented by Mr. Fuller, and proposed to serve as a counter-motion to the recoil of the Armstrong gun—nine of these buffers, about nine inches in circumference and three inches in thickness or substance, were affixed to one of the slides of a gun-carriage employed at the proof, and several rounds were fired, when the gun was satisfactorily and efficiently brought home from the recoil by means of the buffers, and much labor and delay obviated in a succession of firing.

7. Time of Flight of Armstrong Projectiles.—Unlike projectiles shot from cannon of the usual construction, the velocity of the Armstrong projectiles, when shot from the Armstrong rifled pieces, diminishes very little with the range. Their average velocity is about nine hundred and fifty feet per second.

8. Shrapnel or Spherical Case.—Some ingenious improvements have been made in this shell by Lieutenant-Colonel Boxer, of the English artillery. A wrought-iron diaphragm has, at his suggestion, been cast inside it, which is calculated to be of use in preventing the powder from being ignited prematurely by the friction of the balls contained within it. His fuse, too, is more easily applied and more certain in its effects than the old one. These fuses can be prepared in the field to suit all ranges. They should be adapted to burst the shell at about one hundred yards short of the troops aimed at, when the enclosed balls and the fragments of the shell continue to move rapidly onward, and spread sufficiently to produce the effect of a volley of musketry; they are very formidable against columns of troops, even at a range of fifteen hundred yards with field guns, and at a considerably greater range with large pieces.

> The 6-pounder diaphragm shell contains 30 carbine balls.
> The 9-pounder diaphragm shell contains 60 carbine balls.
> The 12-pounder diaphragm shell contains 75 carbine balls.
> The 24-pounder diaphragm shell contains 100 musket balls.
> The 32-pounder diaphragm shell contains 150 musket balls.
> The 68-pounder 8-inch gun contains 300 musket balls.

9. Double-Shotting.—(From the English "Army Hand-book of Field Service.")—*Double-shotting*, which is principally a naval practice, is forbidden with the 10-inch gun and the 8-inch gun of fifty-two hundred weight, and with all carronades.

The greatest effective distances of double-shotted guns, with certainty of penetration through a ship's side, are as follows:

68-pounder, 95 cwt. gun	10 pound charge,	400 yards.
8-inch, 65 cwt. gun	5	"	200	"
8-inch, 60 cwt. gun	5	"	200	"
32-pounder, 56 cwt. gun	6	"	400	"
32-pounder, 50 and 45 cwt. gun	5	"	400	"
32-pounder, 42, 40, 30 cwt. gun	4	"	300	"
32-pounder, 32 and 25 cwt. gun	3	"	200	"

With these charges, the above-named natures of guns may be fired double-shotted, with perfect safety.

The 32-pounder, 56 cwt. gun, may be fired treble-shotted, and produce effective penetration. This is the only gun it is considered safe to fire up to two hundred yards treble-shotted.

In double-shotted guns care should be taken that the two shot be brought into actual contact, for if a space be left between them they will both probably split. The intervention of a junk wad between the two shot renders them more liable to split than if they had been placed in actual contact. The same precautions are necessary when firing a shot and shell together.

10. Penetration.—A 68-pounder solid shot, with 16 lb. charge, will penetrate sound oak timber 45 inches, at 1,200 yards; 32-pounder solid shot, 10 lb. charge, 30 inches; 10-inch shell, 12 lb. charge, 35 inches; 8-inch shell, 5–10 lb. charge, 20–30 inches. The charge for the 8-inch gun should be reduced to 8 lbs. for distances within 600 yards; $5\frac{1}{2}$ lbs. gives sufficient penetration for shells up to 1,200 yards, but the accuracy of fire is, of course, greater with 10 lbs.; 12 lbs. is too much for the shell.

11. Ricochet Fire.—This method was invented by Vauban, and by him styled *Batterie à Ricochet.*

This fire being used for the purpose of reaching objects which can not be affected by direct fire, on account of intervening obstacles; it is employed to great advantage in harassing an enemy when formed, or in the act of forming, behind a rising ground or other obstacle, taking post in a wood, etc., and in enfilading the faces and flanks of a work, a line of troops, etc.

The nature of ordnance best adapted for ricochet fire, are those which throw heavy shells, such as the 10 and 8-inch howitzers, and 68-pounder carronade; for if used to enfilade a work, the shells lodge and explode in the traverses, and render the guns more liable to be dismounted and their detachments put *hors de combat.*

When ricochet fire is had recourse to with field ordnance, the charges and elevation will depend upon circumstances, such as the range, position, and nature of the ground. As a general rule, however, the charge may be reduced by a third or a half less than those for service, and the elevation will be in a corresponding degree increased.

Ricochet fire should never be employed *against a work, if possible,* at a greater distance than six hundred nor less than three hundred and fifty yards.

The elevation above the crest of the parapet required for ricochet fire is from six to nine degrees.

A necessary precaution in commencing ricochet fire is, firing short of rather than over the object, as it is much easier to rectify the first than the second error; by observing the spot which the shot strikes, the additional quantity of powder or elevation required may easily be ascertained.

Ricochet fire was introduced at Philipsburg in 1688, and in 1692 practised at the Siege of Aeth. It was never used in the British artillery, in sieges, until those of the last Peninsular war.

12. Salvo.—A discharge of artillery from a whole battery at once, is called a volley or a salvo.

The fire of *field batteries* should never be carried on *against troops* in salvos, but in a regular manner, well sustained, and with distinct intervals between every round—commencing slowly, and increasing rapidly, as the range diminishes.

When concentrated, the effects of the fire will be in proportion to the number of guns brought together, and therefore, in most cases, in order to strike a decisive blow, this should at once be done.

In *sieges*, concentration is especially requisite to admit of preponderance of fire.

The French, after defining the breach of Ciudad Rodrigo, fired salvos of forty-six guns.

Sir Thomas Graham opened fifty-four guns at 7, A. M., September 7, 1813, against the Castle of St. Sebastian, with such concentrated effect that at 1, P. M., a flag of truce was hung out and the place surrendered.

At Almeida, the French fired salvos of sixty-five guns; the ground trembled, the castle bursting into a thousand pieces, gave vent to a column of smoke, and the whole sunk into a shapeless mass.

The concentration of fire does not, however, necessitate the assembling of enormous batteries *in one line;* it will be sufficient that the guns should be capable of moving, when required, into positions according to the nature of the ground, so as to direct their missiles concentrated on the object upon which it is intended they should take effect.

13. Spiking and Unspiking Guns.—To spike a gun is to drive a long nail or spike so hard into the vent that it can not, without the expenditure of time and labor, be removed. If the end of the nail projecting into the bore be clenched by the rammer, it will further increase the trouble of rendering the piece serviceable again. There are other modes of disabling ordnance, such as breaking off the trunnions, firing a shot into the bore or upon the chase; but if dispatch be required, the spiking will most readily effect the purpose.

Among the implements now supplied to the pioneers of the English army, are gun-spikes. Each pioneer carries two.

The unspiking is accomplished by double-charging and double-shotting the piece, and igniting the powder by means of a loose train of powder to the muzzle, and a leader of quick-match along the bore, setting fire to the quick-match with a short piece of port-fire or slow-match. The operation, if unsuccessful, to be repeated.

In brass guns, a few drops of sulphuric or nitric acid applied more than once, for a day or two, round the top of the spike, will help to loosen it. In the event of failure, a new vent may be drilled half an inch in front of the old one, which can be done in much less time than it would require to drill out the spike.

14. Glass Hand-grenades.—In the Museum of the Royal United Service

Institution there is a glass hand-grenade which was found in the Redan, Sebastopol, on the 10th of September, 1855.

15. Grummet and Junk Wads.—The grummet wad is used in loading with cold shot when the elevation of the gun is under 3°. It is formed by taking a piece of 1½″ or 2″ rope, and making a rough circle rather less in diameter than the bore of the gun it is intended for, fastening two cross pieces or diameters of the same sized rope upon it, these cross pieces projecting a little over the grummet, so that when in the bore of the gun they shall rub or bite upon it, and prevent the shot from moving when the gun is being run up.

Grummet wads are preferable to junk wads for cold shot.

Junk wads are made of old junk beaten into a cylindrical form, and shaped and strengthened by rope yarns passed round them. They are used in firing hot shot.

16. Gun Cotton, or Pyroxyle, was discovered accidentally by Schoenbein, and appeared at first to have so many advantages over gunpowder that it was supposed it would in a great measure supersede it. This substance results from the saturation of clean cotton (or paper prepared without size, and chemically clean) in a mixture of equal parts of concentrated nitric and sulphuric acid, for about a quarter of an hour. It is then squeezed out, well washed in water, and dried carefully, when it is found to have entirely altered its nature, and to have become an explosive material, having about four times the force of gunpowder, weight for weight. There is little or no smoke when gun cotton explodes, and the temperature at which it ignites is so low (356° Fahrenheit) that it may be fired on the top of a small heap of gunpowder without igniting the latter. A heavy blow on an anvil ignites pyroxyle by the heat thus generated, but it can not be fired by friction like the fulminates, and it does not appear to be injured by damp. There are many objections, however, to the use of this material: for artillery, its action is too sudden, and the gases generated injure the metal of the gun, while its bulk and the uncertainty attending its manufacture (two samples seldom corresponding in strength or the facility of ignition), have caused its abandonment by the miner also, and gunpowder still bears the palm among explosive agents. Paroxyline, which is the same as gun cotton, only that the proportions of the acids are slightly altered, is the substance now used so extensively in photography; it is dissolved in alcohol and ether, and it is then called collodion.

The preparation of the explosive cotton is not attended with any danger of explosion, is most simple, and so quick that, within twenty-four hours at the most, the whole operation is gone through and finished. The arrangements for that preparation are equally of the most simple description and little expense.

17. Carcass.—A carcass is a strong spherical shell, pierced with three holes, filled with an extremely powerful composition, which burns with intense power from eight to ten minutes, and the flame issuing from the holes sets fire to everything combustible within its reach; it is consequently used in bombardments, setting fire to shipping, etc., and projected from ordnance like a common shell.

The composition for filling carcasses is as follows—the *proportions* are expressed decimally by weight:

Saltpetre.	Sulphur.	Antimony.	Resin.	Turpentine.	Tallow.
19.0	19.6	6.8	14.7	4.9	4.9

The dry ingredients are to be well mixed; the tallow and turpentine are then to be placed in an iron pot, which fits into a copper containing oil, the temperature of which is raised sufficiently to melt the tallow. The dry ingredients are then to be added, and the whole to be stirred until it is formed into a pasty mass. Corks are then put into two of the holes, and the carcass filled with the composition, which must be well pressed in. When the carcass is full the corks are removed, and wooden plugs forced into the composition nearly to the centre, and allowed to remain there till the composition has set; after which the plugs are removed and the cavities driven with fuse composition, and primed with quick-match. The fuse holes are then covered with canvas patches.

18. Lasso Harness consists of a brown leather surcingle and one trace. The surcingle is rather wider than a common girth, and is composed of two pieces (joined together by rings), one of which is placed over the saddle, and the other round the belly of the horse. There are also rings at the end of the surcingle, which is drawn very firmly round the horse, and fastened tight by lapping a white leather thong (fixed at one end of the surcingle) through the rings. There are two descriptions of traces, one being eight and the other twelve feet long. They have hooks at each end, and, when the lasso harness is made use of by cavalry, etc., to assist draught horses in moving very heavy carriages, or in dragging guns, etc., up steep hills, one of these hooks is fastened to a ring in the surcingle, and the other to the carriage, etc.

Lasso harness may be advantageously employed with all horses; even those unaccustomed by draught having been found perfectly tractable and efficient the first time they were required by means of the lasso. When two horses are in draught, the traces must be inside, and each rider should keep his horse's croup a little outward.

In the Russian army the aid of cavalry, with lasso harness, is made available for rapidly moving and concentrating their heavy batteries of reserve, as was witnessed in their attack upon Eupatoria, during the Crimean campaign.

19. Slow-Matches are prepared from hemp slightly twisted and soaked in a solution of lime-water and saltpetre.

The Gibraltar slow-match consists of sheets of strong blue paper dipped in a solution of eight ounces of saltpetre to a gallon of water, just made to boil over a slow fire, and hung up to dry; when dry, each sheet is divided into two parts, rolled up tight, and the edges pasted down; one half sheet thus made will burn three hours.

20. Ice, three inches thick, will support infantry passing over it in file; four to six inches will bear cavalry and light artillery; and beyond that the passage of heavy guns and carriages over it may be accomplished without danger.

The following mode of strengthening young ice has been suggested: when the water is sufficiently frozen to bear a man, straw six inches thick should be laid across the river, and water poured over it. As soon as the whole mass is frozen together, planks are laid over it. It will bear field artillery. Ice eight inches thick will bear with safety half a ton weight upon a square foot.

21. Passage of Rivers.—A ford, to be passable for infantry, should not exceed three feet in depth. Cavalry can pass a ford four feet deep, but carriages with ammunition can not pass a greater depth than two feet four inches without the risk of wetting the ammunition. The best precaution for crossing a ford is to drive two rows of pickets at certain distances, to mark the best line of direction, and to pass strong ropes along each row, with a turn round each picket.

22. Pontoons.—Blanchard's pontoons are at present the standard pattern in Europe, They are of two kinds. The larger, with hemispherical ends, are 22′ 3″ in length, and 2′ 8″ in diameter; the smaller, with conical ends, 15′ in length, and 1′ 8″ in diameter.

The weight of two of the largest to form a raft is 8 cwt. 7 lb.; weight of the superstructure, 18 cwt. 2 qrs. At the full interval of twenty-four feet from centre to centre of rafts, this bridge will carry infantry four deep, marching at ease, cavalry two deep, and light field guns; at 16′, heavy guns. A raft of three cylinders, at close intervals, will support siege ordnance.

Five of the small or infantry pontoons weigh 5 cwt. 1 qr., 12 lb.; superstructure, 7 cwt. 1 qr. The pontoons at 5′ 4″ from centre to centre, will support infantry three deep, and with care, light field guns or carriages.

Great attention and caution are required in passing over pontoon bridges, the vibratory motion of which is very dangerous, and should be lessened by every possible means.

The troops, in passing, should not preserve an equal pace. There should be no halt on the bridge. As soon as the bridge is perceived to rock, the passage of the troops must be stopped. Cattle should be driven over in very small numbers at a time.

23. Distance by Sound.—To ascertain the distance of an object by the report of fire-arms, bear in mind that the velocity of sound through the air is at the rate of about eleven hundred and twenty-five feet per second, at the temperature of sixty-two degrees; at the freezing temperature it is only about one thousand and ninety feet per second; and, by observing the number of seconds that elapses between the flash and the report of the gun, and multiplying the number by eleven hundred and twenty-five, the product will be the distance in feet, with tolerable accuracy.

Another calculation is, that by multiplying the number of seconds between the flash and the report by eleven hundred, the product will be the distance in feet, with sufficient accuracy for ordinary purposes.

It is obvious that different degrees of temperature, at the several periods of making the calculations, will occasion discrepancies in the results.

24. Tactics of English Artillery.—The English artillery tactics may be sketched in the four following points:

1. Artillery, acting with other troops, always covers the troops when advancing, retiring, or deploying into line.

2. When the line retires by alternate companies, wings, or battalions, the artillery remains with that part of it which is nearest the enemy; retiring with the polonge, and halting when it arrives at the halted part of the line.

3. When the troops are in column, the artillery is on the flank.

4. When the line of troops wheels backward into column, the artillery break into column and close to the reverse flank, so as not to interrupt the line of pivots.

25. Present Artillery Tactics of the French.—The tactics of the French differ from the artillery manœuvres of the English. With the former, the artillery may be said to direct the battle—the infantry conforming, in every instance, to the manœuvres of the artillery—coming up in support wherever the artillery place themselves, without attending in any way to the infantry formations. This principle is one which the Emperor of the French acted upon with singular success throughout the late Italian campaign, and, in doing so, only followed the example of Napoleon I, whose artillery manœuvres will be noticed at length in the succeeding lines.

26. Napoleon's Organization for Offensive Warfare and Plan of Attack.—In adopting a system, as it were, of separate centralization, Napoleon gave to his *corps d'armée* the means of acting by itself. Each had its reserve and park of artillery; and, not having any fixed place in the order of battle, could move independently to accomplish any particular object, and afterward take any special position that might be desirable to enable it to share in the battle. Such an arrangement gave a manifest advantage over the extended order of battle of other Continental armies, since Napoleon could, even with inferior forces, vigorously assume the offensive, and, by breaking through some part of the enemy's line with a superior force, attack him in rear and flank, and also continue this attack against another division, without giving time for the adversary to be aware of the check he had received, or to perceive the danger of his retreat being cut off, by having more hostile troops poured through the interval thus made. A battle of this kind was not general along the whole line, but became a series of engagements at particular points, often distant from one another; and the course followed by Napoleon to obtain a key to the enemy's position, previous to a general attack, appears to have been different to that which has been imagined from the apparent eagerness with which he fought.

Before forming his plan of the intended battle, he allowed that portion of his army which was nearest to the enemy almost to commence a general engagement, in order that the exact position and purpose of the adversary might be ascertained. The reinforcements which were frequently importunately demanded by his generals at this juncture were often withheld, and the temporary success of the enemy was disregarded. But when there was no longer any doubt as to the position and force of the enemy, Napoleon's plan of attack, which was then speedily formed and energetically executed, generally depended upon a great effort to carry one or two points under a preponderating fire of artillery. But instead of employing his whole force, the attack was commenced by a portion of his troops only, which he continued to strengthen until the enemy's reserve had become engaged. The decisive moment having then arrived, Napoleon's reserve was brought up fresh, and having passed through the opening in the enemy's line, one portion attacked him in the flank in rear, while another endeavored to cut off his retreat. It was at this juncture that the whole French army assumed the offensive, and the victory was frequently gained when perhaps three-fourths of the line were ignorant how it had been accomplished.

As the basis of decisive attacks, it was absolutely necessary that the artillery service should unite efficiency to such rapidity of movement as would enable this arm to share in the most difficult and complicated manœuvres of battle; and when thus organized it can rarely fail to secure the success of an attack made on any particular point, or to recover lost ground, if this arm is brought up in a mass to act at a critical moment, such as during the first part of the celebrated action of Marengo. The battle commenced by opening a fire from one hundred pieces of cannon, which proved irresistible; and the French army, being broken and defeated, had already retreated before twenty-seven thousand imperialists under General Melas, when the engagement, then actually lost by the French, was renewed under altered circumstances.

Napoleon, perceiving the approach of Desaix, opportunely brought up fifteen guns under Marmont, which, having opened a destructive fire, checked the advance of an Austrian column, and at this moment, taking advantage of a defile, he first rallied his flying troops, and then became the assailant. This happened at the very moment when some Austrian cavalry had been detached on another service, while the infantry, being then at some distance from the main body of artillery, were actually preparing to avail themselves of the opening supposed to have been made through the enemy, to continue their march to join General Wucassartch on the Adige. Thus the laurels which General Melas had apparently won, were snatched from his brow by the presence of mind and daring intrepidity of his skilful opponent.

27. Table showing the Personnel and Materiel of a Prussian, Austrian, Russian, French, and English Light Battery.

	ARMY OF				
	Prussia.	Austria.	Russia.	France.	England.
Guns	6	6	6	4	5
Howitzers	2	2	2	2	1
Captain, first-class		1	2	1	1
" second-class	4			1	1
First Lieutenant		3	4	1	3
Second "				1	......
Adjutant N. C.	......	......	......	1	......
Surgeon	1	......	1	......	1
Assistant Surgeon	......	......	......	......	
Sergeant-Major			......	1	2
Sergeants	11	15	8	8	4
Pay-Sergeant			......	2	......
Corporals			......	12	6
Artificers or bombardiers	16	5	16	6	6
First-class gunners	114		76	28	127
Second " "		141		54	
First-class drivers	19		48	40	
Second " "				54	
Artificers in wood and iron	......	......	......	4	......
Farriers	......	......	......	3	1
Collar-makers	......	......	......	2	2
Trumpeters	2	2	2	3	1
Fire-workers	......	......	8	......	......
Wagon-masters	......	16	......	......	......
Ammunition wagons	......	24	8	......	14
Baggage wagons	......		3	......	
Spare forge	......	......	......	......	1
Carriage smiths	......	......	......	......	1
Shoeing smiths	......	......	......	......	3
Wheelers	......	......	......	......	2
Horses	222	180	209	268	156
Enfants de troupe	......	......	......	2	......
TOTALS.					
Officers	5	4	7	4	6
Men	162	179	158	218	155
Horses	222	180	209	268	156

CHAPTER III.

FORTIFICATIONS.

How fortifications originated—Their importance and development, with those of
means of attack—What they are for cities, for the country—Principles—Special
character of Vauban—His system upon the frontier of Flanders—Good for Holland—Why—Two kinds of places—Places of depot—Their first condition—Their
situation—Strasbourg—Metz and Lille—Alexandria, Mantua, and Venice—Places
of domination—Places of manœuvres—Where to be established—Mountain
gorges—River frontiers—Importance of shelter for army stores—Detached works
—Fine application—Entrenched camps—Two kinds—Continuous line—Gives bad
results—Why—Morale of an entrenched army—Its duplicity and consequences
—Three examples—Lines of Turin, eighty thousand men against forty thousand
—Lines of Denain, Eugene, and Villars—Lines of Mayence—Affair of October 8,
1795—Lines are good—Against which armies—Eugene of Savoy at Belgrad—
Guaranty of victory in courage, and faults of enemy—Permanent entrenched
camps—Their true value—Towers of Linz—Model not perfectly followed—Their
true usefulness—Fundamental principle—Strategic point well chosen—Importance of their construction—Sudden march upon Vienna become impossible
—Entrenched camp of Verona—Fortifications of Paris—Importance of that
capital—Its fall leads to that of the empire—Part of the detached forts—Paris
not to sustain a siege—Superfluity of enclosure.

NOTES.—1. Marshal Vauban. 2. Strasbourg. 3. Metz. 4. Lille. 5. Alexandria.
6. Venice. 7. Turin. 8. Prince Eugene of Savoy. 9. Denain. 10. Marshal Villars. 11. The City of Paris—Entrenched camp in the Confederate States. 12.
Siege. 13. Time of reducing fortresses. 14. Some practical matters: Hurdles,
gabions, mantlets, parallels, screens of cloth or canvas.

To treat here in detail of fortifications, would not be in accordance
with the spirit of this work, and would probably surpass my strength.
I will then, in this connection, consider the necessities of war, the
proposed object in raising fortifications, and leave aside that which belongs to the art of the engineer.

Formerly fortified places were the results of necessity, and, so to
speak, were formed by themselves. In the times of anarchy, disorders,
and civil wars, of which the Middle Ages offer example, the numerous,
agglomerated, and rich populations were desirous of protecting themselves. They fortified and surrounded themselves with a rampart.
They armed themselves. The means of attack being still in their
infancy, they were thus protected against all assault.

But the discovery of artillery, and the perfection of that arm, soon
changed such a state of affairs. Instead of those ancient places,
valueless against regular means of attack, fortresses built with care and
at the expense of the state, had to be constructed. And since every
place could not be fortified, the government selected those cities which,

on account of their importance, and especially their position, demanded most care and sacrifices. The question was, therefore, no longer the special interests of cities, but, beyond all, the defence of the country against an exterior enemy. Often, however, the determination was owing to accident, without having any sufficient motives for preference.

All great questions should be solved by principles. First, the end must be known and indicated; then the means by which it can be attained will present themselves to the mind; otherwise, we journey at hazard.

In this case, the resolutions arrived at have again been modified by private interests and personal influence, and, it may be said, through the system of war which originated in the epoch of Louis XIV, and which was based upon so many errors.

No one certainly has greater respect for Vauban[1] than myself, but he was more of an engineer than a general; and, in raising many fortresses, he gave himself complaisantly up to labors of his taste.

The number he constructed was, therefore, prodigious. Still, one thing astonishes me in a genius of his nature: it is, that upon an open frontier like that of Flanders, he had the idea to create a barrier of any strength, by means of a system of fortified places, as upon the chessboard.

Had a like system been established for a small country like Holland, the defence of which is mostly based upon natural circumstances which can be profitably employed, nothing would have been better; there, short distances and fortified places, commanding very extended inundations, form great obstacles, add to the means of an army, and facilitate its manœuvres.

But to imitate this system upon an open frontier was an error which ought not to have been committed by a genius of the order of Vauban. If he has not been obliged to conform to superior orders, he certainly gave way, in his character as an engineer, to the attractions and the mania for constructions.

The changes which have taken place in the art of war, and especially as to the strength of armies in the field, have demonstrated the evil of such a system of defence, and no military mind would, nowadays, conceive the idea of constructing again works of such a nature.

Recognized principles have established two kinds of places: places of depot, and places of manœuvres.

The first should be very large, strong, and few in number; one upon a frontier suffices.

They should contain sufficient material for the wants of a large army there assembling, in artillery equipage, in small-arms for necessary changes, and provisions of all sorts. They should have numerous workshops, an arsenal of construction, and, at all times, the material for a large hospital, and subsistence stores. The regiments which are directed there, finally, should leave perfectly organized and armed, so as to be immediately ready to take the field and to combat.

In the next place, reinforcements and replacements of which the army stand in need, are there organized; and, if the opening of the war has been unfortunate, or if the army, inferior to that of the enemy, be reduced from the beginning to the defensive, its forces are doubled by resting upon such a place for support—which, by preference, should be established near a navigable river, to facilitate the arrival of munitions of war. A place of depot thus favoring the movements of an army operating in its neighborhood gives, at the same time, great consistence to its base of operations.

In France we have three places of this kind, wonderfully situated: Strasbourg[2], Metz[3], and Lille[4], for the frontiers of Germany, the Ardennes, and Flanders.

In the times of our greatness we possessed, in Italy, viewing their situation, three places in echelon, which secured its possession: Alexandria[5], Mantua, and Venice[6]. Had our prosperity lasted, another place of great importance would probably have been constructed upon the Save. In newly-conquered countries such places are not only depots for the defence of the frontier; they are, besides, places for the maintenance of the territory which surrounds them.

After places of depot, come places of manœuvre. They serve to facilitate the movements of armies, and to disconcert or prevent those of the enemy.

They should be exclusively situated, either upon rivers, whose two banks they occupy, or on mountains, whose valleys they close.

A chain of mountains presents great obstacles to the movements of an army. Only by the roads which traverse them is there a passage for the considerable material accompanying an army. It is, therefore, useful to close the entrance through these roads by a fortress, so as to prevent the attacking enemy from profiting by them, and thus reserving their use for ourselves.

A river forms the defensive line of an army; an enemy, disposed to clear it, must create means of passage, because the permanent bridges are not in his power. The army placed upon the defensive, on the contrary, can with security manœuvre upon its two banks, and bring all its forces against a part of the enemy's, whenever they are divided. If it happens to engage troops remaining in rear, and which have not yet crossed the stream, those who have cleared the passage are at the power of those sad chances always presented by an isolated situation, without any communication. The most efficacious method, generally, besides the energetic defensive one, consists in offensive movements which are restrained, well calculated, rapidly executed, and begun at the right time.

I shall confine to this the general ideas which should govern the defence of a frontier. As to the details of construction I will only say, in view of the progress in artillery and the facility of its transportation, it is an object which can not too much be recommended to the care of the engineers; without sufficient and perfectly secure shelters for the munitions of all kinds, and a considerable part of the *personnel* of the garrison, there can be no defence possible.

Besides, fortified places should occupy a great extent of ground by means of systematically constructed, detached works, and strong enough to enable each to defend itself. The general defence will be much easier through them, the attack more embarrassing, and the resistance much longer. This kind of fortification has been beautifully applied at Alexandria, in Piedmont, and, if political events had permitted to make it useful, this place would have then rendered the greatest services. But since its considerable extent required a large garrison, and the Piedmontese army being but of medium strength, it did not suit the King of Sardinia to preserve it; the fortifications have, therefore, been destroyed, and are reduced, at this day, to the citadel merely.

I have already, before this, explained what is the aim of erecting strong places, and what are the conditions which should determine their manner of construction and the choice of their locality. I will now speak of fortifications whose object is to cover an inferior army against one which is superior, and to give to the former the faculty of resisting in spite of the disproportion of forces; in short, I speak of entrenched camps, which are destined to establish some kind of equilibrium between forces of different strength.

Entrenched camps are of two kinds. The first kind is composed of one continuous line, which creates material obstacles upon the whole extent of the position which the army occupies; the other kind consists in a fixed number of carefully fortified points, made strong enough, if possible, to be secure from sudden assault. Since they can resist any attack, however spirited, they serve as a support to the troops, protect their flanks, cover a portion of their front, and make them invincible without checking the liberty of their movements.

The first have scarcely ever given good results whenever seriously attacked; they have always been forced. This result is to be attributed to two causes:

In the first place, the troops are too much divided, being obliged to guard the whole extent of the line; often the capture of a single point brings about the evacuation of all others. In the second place, the entrenched army believes itself always inferior, and this opinion robs it of one-half of its valor. If one point of the line be forced, the army no longer thinks of defending itself; although, at this very moment, it would have the greatest chance of victory, because its force is nec-

9*

essarily superior to that of the enemy, who, while penetrating with the head of his column through the breach, prevents the troops which follow from taking any part in the attack; besides, the enemy has to pass a defile. Thus, when it has the best part of the bargain, and should most assuredly succeed, this army thinks of retreating.

Examples of this kind are numerous. I could easily cite very many, but I am content to recall three, which are celebrated, and one of which happened under my own eyes.

The first is the capture of the lines of Turin[7], defended by an army of 80,000 men, attacked by Prince Eugene of Savoy[8], with 40,000 Austrians.

The second happened at Denain[9], where Marshal Villars[10], with an inferior and discouraged army, beat Prince Eugene.

The third is the capture of the lines of Mayence, defended by the French army of 30,000 men, and consisting of works of rare perfection, the most considerable of this kind which have ever been constructed in modern times. Built under the direction of General Chasseloup Laubat, one of the best engineers France has produced, they seemed impregnable. Nevertheless, on the 8th of October, 1795, two small detachments sufficed to create a disorder which nothing could repair; one of 400 men, which passed the Rhine in the rear and up the river, and the other, which was engaged in a small space left between the river and the lines, at the moment when numerous troops in their presence were placed in position for a front attack.

The only reasonable use to construct such lines is to employ them against very numerous but bad armies, such as the Orientals. Their utility in this case has always been demonstrated and recognized; the success obtained by Prince Eugene of Savoy before Belgrad is a new proof. Placed between lines of circumvallation raised against the garrison of the fortress, and lines of countervallation in face of the army of the Grand-Vizier, he was enabled to continue the siege, hold the army in check, take the place, and come out the victor of the strife; but against European armies different principles must be followed.

When a soldier is permitted to place his entire security in a material obstacle in front of him, and when this obstacle is overcome, he thinks no longer of defending himself, and this fatal impression is often communicated to persons of a more elevated grade. A soldier should be convinced, on the contrary, and he can not be too often reminded of it, that the guaranty of victory is, above all, to be found in his own courage, and that he ought to scorn the enemy. But if, instead of having obstacles in his front paralyzing his movements, he finds merely supports covering his flanks and protecting him, he will believe himself invincible, and this impression will soon be communicated to the enemy; and, if resisting an attack, free in his movements, he will know how to profit from a victory, and to develop its consequences.

An army in presence of a stronger army, and under certain circumstances, will then do well to entrench. Supported by a fortified place, a river, or mountains, and surrounded by a greater or less number of defensive points, made likewise as strong as possible, it will thus succeed in supplying the numbers it may lack, and to establish something of an equilibrium.

This subject naturally brings me to the question of permanent entrenched camps—works of recent invention, composed of revetted fortifications, which enclose a large extent of ground, and which, situated upon strategic points, are traversed by a great river. In my opinion, there are no works of more value, and which could render greater services. Several establishments of this kind, although upon different scales and under varied circumstances, are being or have recently been erected.

I shall speak of the two most important, and which have particularly engaged the attention of military men: that of Linz, in Upper Austria, and the fortifications of Paris.

The entrenched camp of Linz is composed of forty-two carefully constructed towers, occupying a circular space of more than six leagues; each one of these towers is casemated, covered from the field side by the relief of the glacis, and its fire is entirely grazing. The model-tower had a deep ditch with a riveted counterscarp, and a gallery for fire in reverse; and I think these means of security in the system have been wrongly suppressed. The armament of each tower consists of a dozen pieces of heavy calibre. All the towers are visible from and sufficiently near to each other to sustain themselves. They occupy, upon a portion of their extent, a range of heights which have, in front of them, in the distance, rugged and difficult mountains, and rest against and are supported by the right bank of the Danube below the city. Upon the left bank a much larger height (the Pessin-Berg) rests against the main branch of the Danube, and is occupied by a suitable and very strong work, from which another line of towers commences, which, enclosing a large space, rests in like manner upon the Danube above the city.

I will not discuss the strength of the isolated towers, believing them capable of but little resistance, were they abandoned to themselves. But covering an army which is enclosed in the space the towers surround, they appear to me incapable of being attacked. The enemy will never undertake a siege, sustained as they are by an army, and the army under their protection need not fear anything.

The fundamental principle of entrenched camps of this kind is the impossibility of being besieged, and to be at the point of the centre of numerous lines of communication. In this respect the camp of Linz is conveniently placed, and its strategic position well chosen.

Two roads, upon the two shores of the Danube, descend this river at a greater or less distance from its borders. Several roads conduct into Bohemia; others lead toward Salzburg, the Tyrol, Styria, and Carinthia. A camp as large as that of Linz, with such obstacles as the country presents, can not be surrounded by the enemy; and the army enclosed therein can never lose all its communications, even if it be supposed that a force triple as large as the besieged army beleaguers it. It will then always be enabled to receive reinforcements and to reorganize itself, until the moment arrives when it is able to assume the offensive; the enemy will then be forced to merely observe—because he will never dare to risk himself into the narrow valley of the Danube, in order to march upon Vienna, as long as an Austrian army remains in this offensive and menacing position.

Truly, such a resolution would be senseless; and if, in 1809, the camp of Linz had existed, Napoleon would not have gone to Vienna; he would have entered it much later. Besides, in war, and especially in the case of large empires, the gain of time is everything; and it is only necessary to give to the natural resources of the country the means to develop themselves. The entrenched camp of Linz is, therefore, an admirable and great military conception.

In every country there are localities which are suited for establishments of this kind, and which would be, when the circumstances arrive, of the greatest utility.

The entrenched camp of Verona is conceived in the same spirit; and, although in its case the circumstances differ, it will, in the hands of a general who knows how to take advantage of it and how to manœuvre, still serve admirably, and ought to play a great part.

I come now to the defensive works erected around Paris[11], which have been and are still the object of long and solemn debates. The construction of the forts, whose system appears to me perfectly conceived, assures a greater independence to France against the attack of the whole of Europe than the acquisition of several provinces, which would have placed the frontier at so great a distance.

No one will gainsay the immense influence which Paris exercises over the destinies of the empire. The disproportionate head of the body, but the active focus, uniting all faculties of intelligence, where an irresistible moral power is developed, where immense treasures are accumulated, and where everything most distinguished in the country finds a common home—Paris has contributed to an extraordinary extent to the power, glory, and splendor of France. But France has to pay dearly for the advantages which Paris confers, whenever the city falls—for with its weight it carries France along. Therefore the interests which influence the entire empire and compromise its existence, must not depend upon the fate of two or three battles; the frontiers must either be extended, or the dangers be diminished which the approach of the enemy brings with it; there is no other means but to prepare an impregnable refuge for the French armies, when, unlucky and beaten, they unite within the walls of the capital.

Whatever may be the consequences of the most fatal campaign, the remains of the army will always be from eighty to one hundred thousand men, and sustained by regularly constructed forts these eighty thousand men will be invincible. Because, with the resources which Paris encloses in *personnel* of all kind, population, riches of every sort, material of every species, and the succors of the neighboring departments, the ranks will soon be filled and all losses repaired; and, in less than one month, an army of three hundred thousand men, well provided and filled with renewed spirit, will march against the enemy. What force will not, then, the enemy need to resist them? If he divides his forces, be will be weak everywhere and easily destroyed; if he remains united to resist and to give battle, how will he subsist? And what will be his fate after the slightest check?

If, then, the enemy has advanced as far as Paris, he can do nothing better than to withdraw, before the moment arrives when the French army, reorganized, will issue forth to seek him; and he must hasten to carry the war into the provinces and within reach of his resources. Then the theatre of war is again upon the frontiers, and everything returning to its natural state, a catastrophe need no longer be feared.

I consider, then, the most auspicious event for the security and defence of France, the construction of detached forts, whose efficiency is such that the enemy can not present himself in force upon many points at the same time. But Paris must not be fortified by a continuous enclosure, because, in my opinion, and in the opinion of all informed and experienced men, this city is not in condition to be able to sustain a siege; it is sufficient to adopt a system of defence which shall prevent its ever being besieged, and with this aim—the only one to be considered—the detached forts are enough; the continuous enclosure is superfluous; and, whatever may happen, the latter can never be usefully applied.

NOTES.

1. Marshal Vauban.—Sebastien Le Prestrede Vauban was born in Burgundy, May 1, 1633, of a noble but impoverished family, eleven of whose members died on the field in the service of their country. At seventeen he began his military life in Condé's regiment, then in the service of Spain—a circumstance little surprising in the bewildering embroilments of that period, when the most illustrious generals of the nation were found fighting now for France, and now for her enemies. He had already been employed as an engineer before he was twenty years old, at which age he was captured by the royalists, and induced by Mazarin to enter the king's service. He speedily rose to his acknowledged position as the first of French engineers, and his military life would embrace a chronicle of all the sieges in the wars on the Flemish frontier—with Spain first, and then with Holland—which filled the last half of the seventeenth century. All the most notable of those sieges he directed in person; but it so chanced that he never was called on to act in the defence of a besieged place. After fifty years of unflagging exertion for the benefit of his country he received the marshal's *bâton* (1703), not then for the first time offered, and died March 30, 1707, full of years and honors, leaving behind him one of the most spotless names in military history, as "The first of engineers and best of citizens;" a noble example in the devoted servant of an absolute monarch—of a patriot in the best sense of the word. Besides carrying the arts of defence and attack to a degree of perfection unknown before, he was eminently a humane and honest man—at once loyal and independent, clear-souled, and guided in the right course as by a natural instinct; in war prodigal of no man's blood but his own. He was often wounded, and a shot in his cheek, which he received at Douay in 1667, marked him to his dying day, and is indicated in his portraits. His exterior was that of a bluff soldier, but his simple and kindly manners agreed with the modest, truthful, and genial character of the man. When Louis XIV, at Cambrai, 1677, threatened a violation of one of the more humane laws of war, Vauban alone fearlessly raised his voice in dissuasion; and not once only, but again and again did this noble soldier, in writing, remonstrate with the king on his treatment of the Protestants and his revocation of the Edict of Nantes. In peace Vauban was as laborious as in war, and left behind him a vast mass of MSS., embracing projects for internal navigation; for the improvement of ports; for the defence of the frontiers, and the fortification of the capital (accomplished in our days); essays on tactics, finance, commerce, the relation between state and church, geography, mathematics, and many other practical subjects besides those immediately connected with his own professional duties. He is said to have been present at one hundred and forty actions; to have conducted fifty-three sieges; to have repaired or improved some three hundred fortresses and ports, besides constructing thirty-three new ones. But, as is well said by the historian of the corps of which Vauban was so long the glory and the patriarch, "Such services as his are to be weighed, not numbered."

2. Strasbourg, on the left, and within a mile from the Rhine, the capital of Alsace, seated near the confluence of the Ill, which traverses it by many arms, now the chief town of the Department of the Lower Rhine, and head-quarters of the fourth military division; having a directory of artillery and of fortification, a school of artillery, a military gymnasium, a cannon foundry, a garrison of infantry, artillery, and pontoon engineers, a park of artillery, etc., and sixty thousand inhabitants. This is the centre of defence of the frontier of the Rhine, and one of the strongest places in Europe; its citadel is composed of five bastions, the outworks of which extend as far as one of the arms of the Rhine, and its system of defence is completed by a sluice, by means of which the surrounding districts can be easily inundated. Opposite Strasbourg, and separated from it by a bridge, stands Kehl, a fortress constructed by Vauban; ceded to Baden by the Treaty of the Ryswick; taken by the French in 1703, 1733, 1793, and 1796; here the French sustained a memorable siege in 1797. The works are now destroyed.

3. Metz is a place of the first class, on the river Moselle, one of the affluents of the Rhine; chief town of the Department of the Moselle, and head-quarters of the third military division; with a directory of artillery and of fortification, schools for the practice of fortification and artillery, and for the construction of gun-carriages and baggage wagons, a school for artillery and engineer artisans, a military gymnasium, arsenal, artillery forges, powder manufactory, etc. There is a railroad from Metz, which joins the Paris and Strasbourg railroad a little to the north of Nancy. This is the centre of defence of France, between the Meuse and the Rhine. The principal works in advance of the body of the place, fortified by Vauban, are the double crown-works of the Moselle and of Belle-Croix, the masterpieces of Cormontaigne.

4. **Lille,** on the river Lys, an affluent of the Scheldt, is a place of the first class; chief town of the Department Du Nord, and head-quarters of the second military division; having a directory of artillery and of fortification, and a garrison of infantry and cavalry; anciently the capital of French Flanders; taken by Louis XIV in 1667, by the Allies in 1708, and vainly besieged by the Austrians in 1792; seventy thousand inhabitants. This city, one of the strongest in Europe, forms, with its ramparts, an oval, the greater axis of which is 7,845 feet, and the smaller axis 3,937 feet; its ramparts have seven gates, and form nine fronts of fortifications—two on the west, covered by the canal of the Deule, and defended by three bastions; three on the north, consisting of seven bastions, crowned by cavaliers, by hornworks, lunettes, etc.; two on the east, covered by three bastions, and the fort St. Sauveur; two on the south, protected by four bastions and the citadel, which is separated from the body of the place by the canal of the Deule, which can be made to inundate the approaches to the place to a distance of upwards of two thousand yards. The citadel is the masterpiece of Vauban; it is a regular pentagon, whose works are so accumulated and disposed that it may be regarded as impregnable. Lille is the centre of defence of the entire northern frontier of France; situated in a plain, it principally covers the low, wet lands, intersected by numerous canals and ditches, which extend between the Scheldt and the Lys. From Lille there is railroad communication with Amiens, Courtray, Ostend, Ghent, Malines, and Antwerp.

5. **Alexandria.**—This city, situated on the right bank of the Tanars, an affluent of the Po, has, upon the left bank, a vast citadel, which is covered by the two rivers; it is the most important place in all Italy, from its eminently strategic position, in the centre of all the roads of the upper basin of the Po. Napoleon had destined it for a grand entrenched camp, by which he would have connected the several fortresses of Turin, Milan, and Mantua; accordingly nothing was spared by him to secure here a safe retreat, a great depot, a formidable rampart for an army, which, having lost a battle, should he be unable to keep the field on the left bank of the Po, and be forced back upon the Appenines; but its fortifications were destroyed by the Austrians in 1814, and there remains now nothing but the citadel.

6. **Venice.**—The river Brenta, after passing the great road from Verona upon Austria, rolls its troubled yet sluggish waters along an artificial channel on a very elevated causeway, turns to the south-east, and terminates, without embouchure, among green and stagnant pools and marshes, where the land ceases imperceptibly and blends with the sea.

It is amid these *lagunes* that Venice is seated on some sixty or eighty islands, and seems to float upon the sea; the canal which separates it from the *terra firma* is about four miles broad, but it has an incessant tendency to be filled up by the earth and sand which are brought into it by the Brenta; and we can foresee a time when this city, the Queen of the Adriatic, will be entirely united to the mainland of Italy. The canals, which everywhere intersect the city and supply the place of streets, are menaced with the same fate, notwithstanding all the care that is taken by the Venetians to get rid of the mud and sands which accumulate in them; and we may say with Chateaubriand that, in a physical as well as a moral sense, "Venice will one day return to the slime out of which she has risen."

The war of 1805 and the Treaty of Presbourg brought about the union of the Venetian states with the kingdom of Italy, when Napoleon, with a view of making Venice a great military station and entrepôt of the commerce of the Levant, ordered the construction of enormous works in the port, and covered the city on the land side with forts and redoubts. The Treaty of 1814 caused her once more to pass under the dominion of Austria.

Her port is the most considerable of the Austrian monarchy, and the seat of all her navy; many new outworks, both upon the lagune islands and in its villages on the mainland, have rendered it a fortress of the first class.

7. Turin, on the left bank of the Po, defended by a citadel, the only relic of its former fortifications; has an arsenal, military school, etc., and one hundred and fifteen thousand inhabitants. Taken by d'Harcourt in 1640; besieged in 1706 by the French; besieged and taken from the French by the Russians in 1799; retaken by the former in 1800, and retained by them until 1814.

8. Prince Eugene of Savoy, celebrated in the contests with Louis XIV of France; born in 1663, at Paris; died at Vienna April 27, 1736, in the service of Austria.

Served in Hungary in 1683; 1691 delivered Coni, took Carmagnole; 1697 commanded the Imperial army; defeated the Turks at Zentha; in War of Succession marched into Italy with thirty thousand men; surprised Corpi, swept the Adige, and beats Marshal Villeroy at Chiari; 1702 surprises Cremona, but fails; defeated by the Duke of Vendôme at Santa Vittoria and Luzzara; defeated in 1704 by the French at Hochstedt; defeated in 1705, in Lombardy, at Cassano, by the Duke of Vendôme; relieves Turin in 1706; gained the Battle of Suza in 1707; wins the Battle of Malplaquet, September 10, 1709, against Marshals De Villars and De Boufflers; 1712, takes Douai; beaten by Villars at Denain, in 1710; in 1717 he fought and gained the Battle of Belgrade against the Turks; in 1733 lost Philipsbourg.

9. Denain, a town on the river Scheldt.

10. Marshal Villars, born in 1653 at Lyons, others say at Moulins de Pierre; died at Turin in 1734. Aide-de-camp to Marshal Bellefons; 1702, beats the Austrians at Freilingen; takes Kehl in 1703, and gains the Battle of Hochstedt; conquers the fanatics of Languedoc in 1704; 1707, breaks the lines of Stollhofen; dangerously wounded at Malplaquet; captures Denain, Marchiennes, Douai, Bouchain, Landau, Fribourg, etc.; commands in Italy in 1733; took Pisighitone, Milan, Novarra, Tortona; marshal-general of the French armies.

11. The City of Paris is divided by the river Seine into two portions, besides the islands. The southern portion is the less considerable and the most elevated; it forms a semicircle, of which the Seine is the diameter. It is protected on the east by the river, gradually disappears to the south on a wide plateau, which, by its blending with the plains of Beance, leaves the city on this side without defence; on the west it may be turned by Saint-Denis, Argenteuil, Saint-Germain, and Versailles. The northern portion is the larger and the more strategically important;

like the other, it forms a semicircle, of which the river is the diameter. It is covered on the west by the Seine from Sèvres to Saint-Denis; on the east by the Marne
from Saint-Maur to Lagny; and, lastly, on the north by a line of heights separating
the waters which fall into the Seine, near Saint-Denis, from those which flow into it
or into the Marne between Saint-Cloud and Lagny. This series of rising grounds, of
little elevation, at first runs along the Marne in undulating hills; it then sinks
down into a plain between between Rosny and Montreuil; rises again in the plateau
of Belleville; is lost in the plain of Saint-Denis; ascends into the high, steep, isolated mound forming the Montmartre; again sinks into the plain of Batignolles; and
terminates in the gentle hills of Chaillot and of Passy, which border the Seine, and
at last disappears in the Bois de Boulogne.

Such is the whole amount of what nature has provided for the defence of the
metropolis of modern civilization; and yet, however inconsiderable these series of
heights may at first sight appear, they offer some military positions—as: 1. To the
north-east, the plateau de Belleville, about 460 feet high, and in extent from 328 to
1,640 yards; it is broken by hollows, and covered with woods, houses, and gardens,
and forms a steep acclivity close to the very walls of Paris, by the mound of Chau-
mont, 377 feet high; Bagnolet and Charonne cover the *débouchés* to the east, Romainville to the north, Pantin and Prés Saint-Gérvais to the west, where it is further protected by the canal of the Ourcq. 2. To the north-west the hill of Montmartre, 427 feet high and 1,100 yards in extent, which looks down upon the walls of
Paris, and is steep on all sides except that toward the city, where the slope is
more gentle; on the east and west sides the quarries that have been hollowed out
of it render it inaccessible, and on the north the village makes it a true redoubt.
It is so strong a position that, protected by artillery, it could never be taken but by
surprise. The heights of Montmartre and of Belleville are separated by the great
depression of the plain of Saint-Denis, an extensive and fertile field without undu-
lations, or trees, or houses, and which is covered on the west by the Seine, on
the east by the canal of Saint-Denis, a derivation from the canal of the Ourcq, and
which enters the Seine close to Saint-Denis, to the south of which it passes. The
plateau of Belleville, the hill of Montmartre, and the plain of Saint-Denis are
thus the military positions which defend Paris on the east and north. Their importance was understood in 1814 by the Allies, who directed all their efforts against
these three points, and here that battle was fought which delivered up the capital of
France to the confederated armies of Europe. A similar disaster is no longer to be
apprehended; the focus of the greatest revolution which has ever happened in the
world is now secured from the attacks of feudal Europe.

Paris is fortified :

1. By a continuous rampart embracing both banks of the Seine, bastioned, and
having an escarpment of masonry of thirty-three feet high; this *enceinte* encloses the
exterior suburbs of Paris, and extends, on the right bank, beyond Bercy, Charonne,
Batignolles, Ternes, Passy, Auteuil, and Point-du-Jour; on the left bank it is carried beyond Vaugirard, Petit-Montrouge, Petit-Gentilly, and Maison-Blanche.

2. By outworks that are casemated, and of which the principal are the forts of
Charenton, Nogent, Rosny, Noisy, Romainville, the works on the canals of the
Ourcq and of Saint-Denis, and the fortifications of Saint-Denis itself, upon the right
bank; the forts of Mont-Valérien, Issy, Vanvres, Montrouge, Bicêtre, and Ivry, on
the left bank.

In the blaze of continued success Napoleon had overlooked the fortifications of
Paris entirely. Nor did the period of his reverses, from 1812—1814, lead him to
protect the capital of his empire. He never dreamed that France proper would

10

ever become, as long as he wielded the power of his legions, the scene of war and devastation. When, finally, in the latter part of the year 1813, his situation became imminently critical, he turned his attention to the capital, but it was too late. Had Paris of 1814 been the Paris of 1864 it would have never been taken.

In the Confederate States we have no such central point as Paris is to France. Richmond, it is true, has become of great importance, no less in a political than in a military point of view, but it neither contains the whole military resources of the Confederacy, nor would its loss dispirit our people as that of Paris did the French in 1814; but Richmond is so far of the utmost importance, that if the casualties of war would wrench it from our grasp, we have not a single fortified point within the Confederacy where we could fall back upon, reorganize, and concentrate at leisure, before an enterprising enemy flushed with success.

It appears, then, to be a matter of the greatest urgency that we should establish one or more *large entrenched camps* where, beforehand, we should concentrate provisions and material of war enough to sustain an army of one hundred thousand men, with fortifications ready, and a haven of refuge at hand, when a day of reverses comes.

There are several very strong and strategically important points both in South Carolina and Georgia, which would make admirable large entrenched camps.

These entrenched camps could serve another useful purpose. The recruits from several states might be concentrated there and instructed *en masse ;* the most important military workshops, etc., could be there united, etc.

12. Siege.—Major-General Twemlow says :

"When the siege of a fortress or strong position is determined on, the first point to be ascertained is the *time* in which it should be taken; the second, the *force requisite* to effect the object; the third, the *reserves* requisite to replace casualties and supplies of all kinds; the fourth, the *strength* of the covering army to prevent reliefs and reinforcements.

"As regards the first and second points, *time* and *force,* they ought to admit of calculation; if means adequate are not available, the siege should not be undertaken; the law of nations should forbid it, in the same manner as a blockade, without a sufficient force to exact it, would not be legal. On the third point, it is clear that, without resources of men, ammunition, stores, food, and requisites of every description, the effective prosecution of a siege is impracticable. And, fourthly, we have the recent example of Mooltan, that a strong place in an enemy's country can not be invested without a covering army preponderant in the field.

13. Time of Reducing Fortresses.—The expense and duration of resistance of a front of fortification of each of the systems, is as follows :

System.	*Probable expense.*	*Duration of resistance.*
Vauban's first system	$200,000	19 days.
" second and third	400,000	29 "
Cormontaigne	300,000	30 "
Coehorn	250,000	21 "
Bousmard and Chasseloup	1,000,000	34 "
Montalembert	1,500,000	80 "
Carnot	500,000	18 "
Mr. Ferguson's from	40,000 without casemates,	
To	300,000 with them.	

14. Some Practical Matters.—Bursting open Gates.—The simplest method is to suspend a bag of gunpowder, containing fifty to sixty pounds, near the middle of the gate, upon a nail or gimlet, having a small piece of port-fire or Bickford's fuse inserted in the bottom. Leathern bags are best for this purpose, but sand bags filled with powder, propped up and ignited, will demolish almost any gate or barrier.

If, instead of being suspended, the powder should be placed at the bottom of the gate, any spare time might be advantageously employed in heaping rubbish, stones, or any other available heavy material, over it, as completely as time and circumstances may admit.

Hurdles were much used by the ancients in their field-works, and are still occasionally found serviceable for revetting, or for laying on wet ground alternately with beds of fascines. Layers of hurdles, covered with heath and ballast, were extensively used by Stephenson in the substructure of the Liverpool and Manchester railroad where it crosses the Chat Moss. In the trenches at Antwerp, deluged with incessant rain, the French laid double tiers of fascines, and over these a layer of strong hurdles, for the passage of artillery.

Hurdles for portability may be made six feet long by two feet nine inches broad, weighing about fifty pounds when dry.

Revetments may be made by hurdle-work, by driving stout sticks along the face of the slope to be revetted to a depth of about two feet in the ground at its base. Branches are then woven in and out between these stakes, and vertical binders applied, when the wattling is completed. Similar wattle-work is employed in Flanders for revetting the submerged escarps of wet ditches.

Gabions.—Up to the year 1853 they were made of wicker-work, but in that year it was proposed to make them of plain sheet-iron, when they were only required for a temporary purpose, and of galvanized iron when required for permanent use. They have since been made of various shapes and in various ways. At Sebastopol, where there was a great want of them, the hay-bands were randed, as it is called, round upright pickets, and they were used with good effect.

More recently a gabion was prepared by Sergeant-Major J. Jones, of the English Royal Engineers, constructed with wrought-iron bands and twelve upright pickets, which having been most favorably reported on by the Royal Engineer Permanent Committee at Chatham, and the Ordnance Select Committee, Sir John Burgoyne, I.G.F., in October, 1860, gave directions for the invention to be adopted generally in the service, and included in the list of stores to accompany armies in the field. The new gabion is formed of bands of common or galvanized sheet-iron, known as twenty-inch gauge, three and a half inches in width, fixed on wooden pickets. The advantage it possesses over the old kind of wicker gabion, hitherto in use by the royal engineers in the construction of their earth-works and defences, are of the most striking character. The old description of gabion occupies three men three hours in making; whereas, on a recent occasion, in the presence of his Royal Highness the Duke of Cambridge and staff, two sappers made one of the iron gabions in four and a half minutes. From experiments made, it has been ascertained that one hundred men can make five thousand four hundred of these iron band gabions in nine hours; while the same number of men in the same time would only make one hundred wicker ones. The chief merits of Sergeant-Major Jones' gabions are that they are more portable, inasmuch as one hundred of his gabions require less room for stowage than six of the wicker ones; and they are much lighter, each weighing about twenty-nine pounds, or thirty-one pounds less than the old kind; they are much cheaper, costing complete 5s. 1d. each, or 1s. 1d. less than the present description; more simple in their construction, and more durable. Being of iron, they are, of course, incombustible; and the bands are applicable to the construction of flying suspension-bridges, hospital beds, ambulance litters, stabling and hutting for cavalry and infantry troops on active service.

MANTLETS.—Captain Tyler, English Royal Engineers, 'in an admirable lecture, delivered on the 16th of April, 1858, before the Council of the United Service Institution, recommends the use of mantlets in sapping operations. "This method would," he says, "have the great advantage of enabling the workmen to place several gabions at a time instead of one only; of enabling them to work in larger numbers and in greater security; and, what is more important than all, of enabling them to carry the trench forward with much greater rapidity. These mantlets need not be very portable, nor need they be capable of being moved rapidly; all that is necessary is that they should be musket-proof, about six feet high by two feet five inches broad; should be placed on wheels, and should be movable slowly in any desired direction, each by one man. They ought also to be furnished each with a couple of loop-holes for the purpose of observation, as well as to enable its occupant to fire, when necessary, toward his front, without exposing himself."

A writer in Blackwood's Magazine, December, 1859, in giving an account of the bloody and disastrous fight in the Peiho river, June 25, 1859, alludes to a striking innovation in Chinese warfare, by an ingenious and successful application of mantlets in the defence of the Takeu forts, and adds:

"These mantlets would be quite worthy of imitation in our own fortifications, and the cleverness with which they are worked deserves all praise. Had they been fitted to the upper port or embrasure-sill, any accident to the lanyard would have caused them to fall down and block up the gun-port, so that they would have to be blown away to enable the gun to work; but placed as they were, by attaching the lanyards to the gun-carriage, as the piece recoiled it closed its own mantlet, and if the lines were shot away the mantlet merely fell down, and left the gun to fight in an ordinary embrasure.

"They were of stout wood, covered externally with a wattling of rattans, so as to be rifle-proof. The mantlet worked on hinges, or rollers, fitted to the outer and lower edge of the embrasures, and was triced up and lowered down by means of lines leading upward through the parapet on each side of the gun. When closed up, the casemated embrasures were not easily detected in the smoke of action, and the gun was loaded and laid point-blank before being run out. Directly all was ready, down went the mantlet, out ran the gun, a shot was fired into the mass of vessels, and as the gun recoiled the mantlet went up again with such expedition that our men required sharp eyes to detect which of the enemy's embrasures was firing, and ought next to be silenced."

PARALLELS.—The distance of the first parallel varies according to the range of grape and musketry. At Sebastopol the first parallel was constructed at three times the usual distance from the works, namely, 1,800 yards.

SCREENS OF CLOTH OR CANVAS.—"Screens might be so useful on many occasions of both attack and defence" of fortresses, "that it is surprising they have not been oftener employed." Sir J. Jones mentions somewhere, that "At Badajos the British engineers extended a canvas screen to cover an unfinished boyeau; and the French, mistaking it for an earthen parapet, suffered the excavation to be completed without molestation. Similar screens were used at Gibraltar," during the siege in 1781, "to mask a thorough repair of the batteries overlooking the neutral ground. They are recommended also by Albert Durer and Maggi, in their treatises on fortification."

The paragraph above, marked with inverted commas, is an extract from Lieut. Yule's work on Fortification, published in 1851. Albert Durer, to whose treatise it refers, and who recommends the use of screens, published a book on fortification in the year 1527 ; and Maggi's writings on the same subject were printed in the sixteenth century. So that the application of such a mode of shelter or concealment in the attack or defence of fortresses is known to be of early date. Yet, on one occasion, a reference to its utility being made in an assemblage chiefly composed of naval and military officers, seems to have excited a little merriment.

Mr. Ferguson, in his "Portsmouth Protected," says : "One of the objections made to my system was, that it would be easy to place a few riflemen in pits or in the nearest parallel, and that they could easily keep down the fire of the place by killing any man who ventured to approach the guns. There did not appear to me any difficulty in avoiding this danger, and, consequently, when the discussion at the Royal United Service Institution came on, I took with me a few yards of baize, with two or three iron rods, and showed how I would propose to stick them along the crest of the parapet, and to hook on the green baize, letting it drop down across the embrasures in front of the guns. The baize was full of slits, through which the besieged could see or fire without the possibility of their being seen by the besiegers. This unorthodox expedient was received with 'great laughter,' and afforded an excellent opportunity for criticism during the following three nights. When we entered Sebastopol we found curtains of rope hung up, exactly as I proposed, across the embrasure, and rope wound round the guns so as to stop the hole in the curtain through which the gun was fired. This rope-cloth, if I may so call it, was rifle-bullet proof, and would probably stop grape, while round shot would pass through without doing it much injury. There can be no doubt that the maintenance of the fire till so late a period was mainly owing to this expedient. The question is, whether was the Russian expedient or mine the best? Theirs was expensive and cumbersome, difficult to apply in all circumstances, and if once damaged not easy to repair; mine was light and cheap, available everywhere, and replaced in a moment if knocked over. It is true it would not stop a ball; but blind fire from rifles is a very innocent amusement, and as the parties fired at were invisible there was but little to fear from this cause. Experience only can decide which modification was the best, but with all due deference to the Russians, I am inclined to think the lighter mode will be found most generally applicable."

On the occasion of the English assault of the works on the Peiho, the embrasures and guns of the Chinese batteries were effectually screened by mantlets and matting from the observation of the attacking gunboats, until the opening of a well-directed and most destructive fire disclosed them.

Captain Tyler, English Royal Engineers, reviving the subject of screens or curtains, in a lecture at the Royal United Service Institution in April, 1858, says : "It is evident that mere screens of canvas, or other suitable material, would be of great use in temporarily obstructing the view of the besieged, and hiding from them the movements and projects of the besiegers for a sufficient time to enable the latter to throw up parapets of a more permanent nature. Such temporary screens might be made to cover a considerable space, very much greater than that required for the operations of the besiegers, and, like false attacks, they might be erected for the purpose of misleading the garrison, in places where they would serve no other object."

10*

CHAPTER IV.

ADMINISTRATION.

Its true base the legitimacy of consumption—Confusion under Directory—One
hundred and fifty thousand men not existing—Corps should provide for themselves—Responsibility of chiefs—Messes of economy—Their importance.

Men congregated in bodies have wants; the talent to satisfy them
with order, economy, and intelligence, forms the science of administration.

The basis of a good administration is the care bestowed upon the
economy and lawfulness of the consumption of army stores of all kinds.
Wherever inspections are thorough, and where the effective and the present for duty are stated precisely and frequently, we find the elements of
order; because great abuses less often consist in an increased cost of
consumed articles, than in consumptions which have not taken place,
and which are yet charged for.

In the times of the Directory, the French military administration
was in a great state of confusion, and the First Consul hastened, upon
his accession to power, to create a new corps, charged with inspections,
to establish order.

He gave to this corps an exalted position, which was justified by great
zeal on its part. At the expiration of six months more than one hundred and fifty thousand men who did not exist, but for the greater number of whom provisions, pay, and clothing were issued, were struck from
the rolls.

The administrative system varies with the countries; all are susceptible of good results, when the effective and the present under arms are
precisely stated. I will only observe that (in my opinion, at least)
great advantages are derived from giving to troops the liberty of supplying *themselves* as much as possible. Since the efficiency of troops
always depends upon a good administration, chiefs of corps should not
only be responsible to a great extent, but should also be invested with
great powers; their operations should be watched, but they ought to
have the direction. If soldiers know that their commander is charged
alone with the responsiblility, their zeal will be better guaranteed. Colonels who transgress must be exemplarily punished, but the glory of success should alone belong to them.

The formation of economical messes in corps has been forbidden in
France, and a profound error thereby been committed. The congrega-
tion of soldiers in messes has always its advantages, and the skilful
and intelligent chief of a corps can and must always insist upon economy, without depriving his soldiers of the enjoyment of any of their

rights. If they are proscribed, they will not the less be formed ; and not being openly avowed, the continued formation of them bears a culpable and mysterious character. If, on the contrary, they are not only authorized, but even ordered, and left to the disposition of the chief of corps to institute them to the advantage of the regiment, in accidental cases and when special orders have not provided for the troops, they will be very encouraging ; and the colonels will feel proud of the success of an institution which yields honorable fruits to them.

Two very important branches of administration are faulty in almost every one of European armies—hospitals and subsistence. An enlightened government should seek to establish them upon a new basis. Important and direct advantages, influencing the art of war, the welfare and conservation of soldiers, would result therefrom. I shall first consider the subsistence of troops.

FIRST SECTION.

SUBSISTENCE.

Bread—Greatest obstacle of war—Problem solved by Romans—Easy then—Why more difficult to-day—Efficacious method—Soldier able to provide for himself—Happy experience of this method—In what great difficulty consists—Able to live with flour—Not with the wheat—Scattering of labor—Soldier knows how to make his soup, if he has the materials—Should make his bread also—How—Army of Portugal—Napoleon employs same means—Portable mills—Their condition—Objection—Is without force—Bran in the bread—Facility of method for administration—Question of nourishment solved—Extemporaneous ovens—Magazines of wheat in times of peace.

In treating of the food for troops I will only speak of the supplying of bread; it alone presents difficulties—the requisite living stock being almost continually within reach of the consumers.

The difficulty of distributing bread regularly to troops is one of the greatest embarrassments in war. It is inexplicable that so many distinguished generals, who, from that cause, have seen their projects delayed or miscarried, should not have solved so important a problem.

The Romans had solved it; but, generally, their wars did not require as rapid movements as in modern warfare.

There is, I believe, a perfectly satisfactory manner to conquer this difficulty, and the change I propose would powerfully influence the art of war.

To permit the administration of subsistence to supply the distribution of bread regularly, the army must be either stationary or retreating, and either always remaining within the same distance of its magazines or approaching toward them. If marching in advance, steadily withdrawing from them, any commissary, however skilful he be, will find the supplying impracticable; and, because the trains can not go faster than the army, they will always follow at the same distance as when departing from the depots; with each new journey the distance augments and the difficulty becomes greater.

In wars of invasion troops can only subsist upon the resources of the country through which they march. But the time necessary to make bread when arriving in inhabited places, which are ordinarily not supplied with a sufficient number of mills and bake-ovens, or their distance, render local resources very incomplete, and the scarcity resulting from it leads to great suffering and disorders. And the maintenance of order in every body, and under all circumstances, is the safety of armies.

The only efficient mode of ensuring the regular subsistence to the soldier is to charge him with supplying himself, according to a fixed rule. I made this experiment, and the result was completely favorable.

War is not made in a desert, and when this passing circumstance happens, dispositions are made for it. War is ordinarily waged in inhabited countries, and wherever men are, there is grain for their subsistence. The manner of using this grain, with which the granaries are filled, appears, then, the solution of the question.

The great difficulty is to reduce grain into flour, as I will hereafter explain. Mills are requisite to grind the wheat; when necessary, soldiers can live on flour alone without converting it into bread, but they would die of hunger on grains of wheat.

When laboring hands are scarce and dear, powerful machines can with advantage be used in manufactories, and labor is centralized; but when labor is superabundant and costs nothing, a system absolutely opposite must be followed. When labor is divided among a great many it is made easier, and by confiding it to those who profit therefrom, their zeal and punctuality will be assured. Thus considered, it is evident that the hands of the soldier can be employed without any drawback, and they will gain by it, in receiving, as an indemnity, the amount which is expended to have the work done by others.

Why do soldiers never fail to have soup in the field, if they are supplied with meat, bread, and a kettle? Because they themselves cook it. If a commissary, under whatever pretext, would take this task upon himself for a division, or even a colonel for his regiment, soldiers, when marching, would never have anything to eat.

I design to apply to the bread the example of the soup, and the soldier will never want for it. I propose to give to the army portable mills; I employed this measure during a campaign in Spain, and it completely succeeded. The Army of Portugal, in 1812, thus lived for six months; the only inconvenience experienced was the rapid wearing out of the mill-stones; they were replaced by those of a better quality, and lasted very long.

Napoleon, being made aware of these results, was struck with the advantages obtained in the midst of the miseries of the Russian campaign, and he ordered a large number of these mills for the Grand Army. Five hundred were sent to him, and they arrived at Smolensk at the same time with the army returning from Moscow. But then he had neither hands to work them, nor soldiers to use them.

The following qualities should be united in these mills:

1. Light enough to be carried by one soldier, who is detached from the ranks for that purpose, on account of the importance of the matter, as soon as the regular means of transport are wanting.

2. To be worked by a single man.

3. To give fine flour, and a sufficient quantity, after a work of four hours, to supply the wants of a company.

The mills of the Army of Portugal gave thirty pounds of fine flour per hour. This system has been objected to, because the regulations prescribing the extraction of the bran, this operation complicated the fabrication of the flour. I reply that experience has taught the inutility of the extraction of the bran, whenever wheat of good quality was used.

With second-class wheat, but pure and without mixture, the bread is still good. When the administration issues bad bread, the soldier must necessarily accept and eat it, under pain of dying of hunger, because the necessity of consuming it is immediate; but when the wheat which has been distributed is full of dust and mixed, it can be cleaned before being used, and the soldier will still have good bread to eat. Thus, in this respect, his condition will be improved; it would be still more so by the payment for the work he has done, be it in money, or in the augmentation of his ration.

But let us consider what effect this habitual simplification will have, in times of war, upon the administration of subsistence, and the facility of its service. A general-in-chief, nowadays, makes greater efforts of mind to assure the subsistence of his troops than he does for anything else, and unceasingly his combinations are interfered with and destroyed, because the bread has not been distributed at the right time.

Thus not only the question of indispensable nourishment of troops has been solved, but also that relating to bread itself. Means can always be found to make a simple excavation in all kinds of ground, and in four hours for the construction of ovens, which, two hours afterward, can be used for baking bread. In each bivouac, then, sufficient flour can be made for daily consumption, and at each daily halt ovens can be constructed near any farmer's house to bake bread in advance. From that moment an army provides for its own subsistence, the administration is no longer occupied with those details which secure to each soldier the circulation of his blood; it is the consequence of a principle always in force.

In times of peace the government should have magazines of wheat, to be distributed to the troops. In a defensive war it should be the same. In a war of invasion, each regiment would daily receive the necessary wheat from either the administration of the invaded countries, or would take it from the granaries of the inhabitants. But this custom should be followed and contracted during times of peace, because, in principle, the usages of peace should be, as much as possible, assimilated to those of war; and this truth is especially incontestable when the question is the introduction of some great change.

SECOND SECTION.

HOSPITALS.

Inherent spirit of warrior—Duty of conscience and humanity—Duty of interest——Pecuniary interest not sufficient recompense—Morality—Recompenses of the sentiment—Hospital nurses—Paid individuals—Spirit of charity—Hierarchy to establish—Service in hospitals entrusted to three corps—Medical science—Material of hospitals—Service of nurses—Knights of Malta—Causes of their change—Lively prepossession—Impracticable idea under Restoration—Why—Utility in Army of Africa—Importance of objections—Vanity of ridicule—Ameliorations—Present hospitals—Hospitals should be within reach of troops—Inconvenience of long transports—Aggravation of maladies—Encumbrance—Proof accomplished.

NOTE.—Chisolm's Surgery.

Nothing is more sad in armies than the frequent spectacle which military hospitals present. A sufficient degree of attention is scarcely ever, in such places, bestowed upon a class of men who, nevertheless, are entitled to the right of universal solicitude. A life of devotion is their existence; sufferings, fatigues, and dangers, are their only prospect. The noblest sentiments animate their heart, and these generous men ask their chiefs only to love them, and to be just in the exercise of their authority. Such is the inherent spirit of the warrior, and it belongs particularly to the French soldier, who is a stranger to none of the sentiments which honor humanity.

There undoubtedly exist vices and bad passions in armies, as in all unions of men; but the example of the highest virtues is likewise to be found. The conservation of sick and wounded soldiers is, then, a duty of conscience and of humanity. It is at the same time of great importance for the government, as for the general, because the largest number of soldiers is an element of success, and their replacement by recruits, very dear in itself, is very far from making the loss good.

What confidence, besides, and what energy, does the certainty impart to a good soldier upon the battle-field that, in case of being wounded, the most efficient succor will be lavished upon him.

With this object, perhaps, it would be well to try to change the spirit of hospital administration; to seek a more noble recompense than pecuniary interest; to develop thoughts, worthier and more elevated, in order to sustain courage and devotion the more.

If the functions of those who administer to the wants of the sick and wounded were raised, ennobled, and compensated by public opinion, and the enjoyment of the powers to enable them to exercise charity and to call forth the sentiments of piety, the result would be of the greatest benefit to those who suffer. To attain this end the care of military hospitals should be left to a religious corps, acquainted with the lower functions of surgery and medicine; which, although not charged with the administration proper, and the management of funds, should have the sole attendance of the sick.

A corps of hospital brothers, engaged for lifetime, or a fixed period, having honored chiefs, should be charged with the keeping of the halls, and with the service near the sick. Paid assistants should be placed under their orders, and employed in the lowest and most disagreeable duties; but no chief should consider any duty, however unpleasant, to be beneath him, in cases of necessity. The spirit of charity would sustain them in their labors. A detachment of these respectable brothers, after having been assigned their destination, ought never to quit those confided to their care. Their presence would be the hope and consolation of the sick; and their holy ministry, exercised over all, friends or enemies, would be their safeguard with all European armies, should the fate of arms bring them into hostile hands.

The rewards of conscience should be their principal recompense. A wisely combined military government should require that blind obedience be accorded to this corps, devoted, as it would be, to the practice of the most touching virtues. The general of an army should sometimes receive at his table, and place upon a seat of honor, the superior of the hospital brothers; he would thus honor all his subalterns, and pay them with that precious money, the value of which increases in proportion to the liberality with which it is bestowed.

Hospital service would, then, be divided into three corps:

1. Medical men, doctors, and surgeons.

2. The administration, providing the material, expending the funds, and furnishing subsistence.

3. Hospital brothers, whose duty is to administer to the wants of the sick, and to see to the application of all proper means to restore the suffering soldiers to health.

This last corps would somewhat exercise a controlling influence, energetic and always watchful, to guard against administrative abuses.   It would be the guaranty of order and regularity.

The care bestowed upon the pilgrims to Jerusalem was the origin of the Knights of Malta, and charity was their first virtue.   The anarchy and the disorders which reigned in the places where they had established themselves obliged them to be armed for their defence, and thus, while remaining hospitallers, they became soldiers also.

Courageous conduct and the profession of arms have ever, and will always, continue to please the multitude, and with reason.   The rôle of soldier having finally engrossed their attention entirely, the character of their hospitality became changed.   The creation of their order resulted from the necessities of a certain class of society; the order which I would wish to establish could not fail to better the condition of a class of men which I consider to be worthy of the greatest interest, and who constitute, in Europe, an energetic and truly patriotic part of every nation.

It would not be difficult to prescribe regulations for hospital service; but to do so here would be beyond the limits of my work.

For a long time past, under the Empire, and when witnessing disorders of this kind, I have entertained this idea.   Its execution was not practicable under the Restoration; an unworthy construction might have been put upon it; but the moment has come, perhaps, for its successful and useful execution.   Of what support it would have been to the Army of Africa!

I do not conceal the objections which could be made to an establishment of this kind, nor the difficulties of maintaining sufficient harmony between three rival corps having the same object in view; but two of them very often already may be very far from any perfect understanding with each other, and a third corps, without adding much to the complications, could not but be useful in enlightening the others.

I am aware that some will not hesitate to ridicule this institution; but I willingly brave it, knowing that it would contribute to ameliorate the condition of the soldier—in my opinion, a matter of the greatest importance for the service as well as for humanity.

In the last years the hospital service has, however, been improved by the military organization of the employees.   An order of advancement, such as with troops, gives a future to those who serve well, and is a means to increase both surveillance and discipline; it calls forth sentiments of honor which make the exercise of authority more easy.   Its results should, then, be good.

Military organization generally ensures at all times the regular exercise of power; it constitutes, then, essentially a great means of order; it will always be employed with success in the case of a union of men laboring to attain the same end; and the greater the confusion, the greater will be the profit and advantage resulting from its employment.

One more word upon hospitals.

Miserable and illusory calculations of economy in regard to what is called the " daily expenditure of hospitals " often restrain, in too great a degree, the number of these establishments; and the desire to charge other persons with our duties renders evacuations of hospitals too frequent. Nothing is more sad than these two systems, when they are not dictated by imperious circumstances—such as the neighborhood of the enemy, or the absolute want of means. In ordinary cases, hospitals can not be placed too near the troops, nor can the sick be too much divided. Generally, simple diseases are cured in a few days when at once treated. They are aggravated when the sick have to be transported too great a distance; and long return voyages, after the illness, exhaust men still weak, and produce relapses which another voyage renders fatal. Thus, in multiplying hospitals and by placing them within the reach of troops, the cure is facilitated, maladies are secure from aggravation, and the sick from weakness; and that great accumulation is prevented which leads to contagious diseases, the terrible sources of the greatest ravages.

Such a system appears to require a great expenditure of money, but in reality it is attended with much economy.

I have constantly followed it, and the troops under my orders have always been in fine condition.

NOTE.

Chisolm's Surgery.—The officer who aspires to the honor of discharging his duties to the fullest extent, and to watch over the welfare of those he commands in every particular, will not consider it the least of his charges to see that his men are properly provided for when well, and to look after them when they are sick.

To give to him a consciousness of the momentous trust reposed in him, and to bring matters before his consideration which thousands of officers are only too apt to look upon as trivial, every regimental officer is earnestly requested to read and ponder over the admirable work of Major J. J. Chisolm, Surgeon C. S. A., on " *Military Surgery*," the third edition of which has just been issued by Evans & Cogswell, Columbia, S. C.

A more valuable and popularly written work on the important subject it treats can not be found in any language.

11

CHAPTER V.

MILITARY JUSTICE, AND COMPOSITION OF COURTS.

Army a particular society—Character of military justice—Should be given to whom—Military judges during Revolution—Error seen—Creation of councils of war—Military justice not absolute—Not founded upon same principles as civil justice—Here regiment forms the unit—Colonel—Regimental courts—Motives of legislator in regard to courts—By division—Composition of councils of war—All grades represented—Why—Indulgence rather with superior grades—Right of grace—Should be reserved to colonel—Usage in Austria.

NOTES.—Articles of War of the Confederate States Army. Imperfections. French code of military punishments. Remarks on our present system of courts-martial.

Social order can not exist if the conditions of its existence are not fulfilled. It is the same with an army, which presents the example of a particular society, governed by special regulations and customs. To discover the principle which solves the question of military justice, we seek the definition of the latter, and find it to be the accomplishment of disciplinary measures. To whose hands should the execution of military justice be confided? To the hands of those who are engaged in the maintenance of discipline, who every day feel its necessity when discharging their duties, and who are most interested in it. To the officers in active service this charge should, then, be exclusively entrusted.

However, it has not always been thus. During the Revolution military judges, who were civilians, and who accompanied the army, were appointed.   The error which had been committed was soon perceived; the saddest consequences ensued, and councils of war, such as they now exist, were created.

In 1829 this matter was again agitated, and a new law of military justice was laid before the Chamber of Peers.

A commission, composed of men of eminent merit, but strangers to the knowledge of troops, proposed to substitute for temporary councils of war permanent tribunals of war, presided over by general officers. This new mode, in establishing a military magistracy distinct from the army proper, would have had all the drawbacks of the system temporarily adopted under the Republic, and would have furthermore lowered, in the eyes of the troops, the character of generals, who are essentially men of combat—who should, by their presence, awaken ideas of glory and rewards, rather than thoughts of crime and chastisement.

Military justice is not established, in any absolute manner, upon moral principles; it has for its basis necessity.

Undoubtedly, in the opinion of every sensible man, when considering morality and personal security, there is a wide difference between the thief and the soldier who disobeys his chief and insults him in a passionate moment. Yet the punishment of the military man will be much more grave; to vindicate society it will be sufficient, in most circumstances, that the one should go to the galleys, while the army would be lost if the other were not put to death; because, in that moment, all bands would be broken, and the military edifice, which is based upon respect and submission alone, would crumble to pieces without this sustaining pillar.

There is, then, an immense difference between civil and military justice. The latter appears barbarous, but it is indispensable; and its execution can only be guaranteed by those who themselves are directly interested in its proper existence.

If the battalion forms the unit of combat for troops, the regiment forms the social military unit, the family, and the clan. The colonel, the chief of this society, is invested with a sort of magistracy, which should watch over its conservation.

He should punish; *he* should assure to every one prompt and impartial justice, and maintain daily order, and the observance of the laws upon which this order reposes. When armies were regularly formed, each regiment had its own court, under the high surveillance of its colonel; and even at that epoch this was not only a necessity, but a right—because every colonel, being the organizing power of his regiment, was obliged to have legal and extensive privileges, which were the guarantee of the obedience of his subordinates.

Regimental courts still exist in several of the great European armies. Placed within immediate reach of those amenable to law, their action can always be made to be felt without any delay. This consideration is of such capital importance that it perhaps should be preferred to the French and Russian systems, where courts are only established by division.

The motive which has influenced the legislator is easily comprehended: he wished to place those accused beyond the personal passions of the chiefs, by trying the former by a court composed, in the greatest part, of officers not belonging to their corps. On the other hand, these officers being in active service, employed with the troops, one is assured that their verdict, rendered without prejudice, will have that degree of severity most capable of furthering the good of the service—because the presiding colonel will act so in the interest of another regiment as he would wish the colonel of that regiment to act at other times in the case of his own regiment. The interest of the army will always be considered.

Every council of war has been composed of members of different rank. This is a compliment paid to the sentiments of duty which equally pervade all classes of military organizations, and a guarantee to the accused, who thus will have one or two of equal rank with himself among the judges. A composition of this kind has no danger, because indulgence, should any be feared, is rather with the more elevated than the inferior grades.

I conclude, then, with the observation that all military courts should be exclusively composed of individuals in active service, and even belonging to the corps placed under their jurisdiction.

Another feature would, perhaps, be desirable in military justice. It exists in Austria, and the effects appear to me salutary. The right of grace, and the commutation of the sentence, is not reserved to the sovereign; it belongs to the colonel-proprietor of the regiment, who, according to usage, delegates the exercise of this power to the colonel commanding. There are so many circumstances which may be found in favor of a soldier guilty of breach of discipline (and it is generally always in such cases that the power of grace is interposed), and the chiefs, placed upon the spot, are so much more apt to appreciate and determine the propriety of an act of clemency, that it appears to me very useful to give this prerogative, not to the chief of corps, but to the general commanding the division or the corps d'armée.

In the actual state of things a brave soldier, whom every one would wish to save, perishes, the victim of the rigor of the law; or, on the other hand, to one who has justly forfeited his life in the interest of his conservation, justice is not dealt out—an equally grievous alternative.

NOTES.

Articles of War of the Confederate States Army—Imperfections. French code of military punishments. Remarks on our present system of courts-martial.

A review of the Articles of War will disclose a very important imperfection, viz: the very great discretion given to courts-martial in determining the punishments for military crimes and misdemeanors. An analysis of the Articles of War will show that there are definite and determined punishments in but twenty-one cases, and but *four* different punishments. They are:

Art.	*Crime or Misdemeanor.*	*Punishment.*
2.	Improper behavior at places of public worship............	Arrest and fine.
3.	Use of oath, etc...	"
14.	Signing false certificate...	Cashiered or dismissed.
15.	False muster........	" "
16.	Accepting bribes..	" "
17.	Mustering civilian as soldier..	" "
18.	False return..	" "
22.	Harboring or receiving deserters.................................	" "
25. 26. 28.	Principals and accessor to duels............................ Upbraiding for refusing to fight duel.........................	" " " "

29. Sutler breaking regulations ....Cashiered or dismissed.
31. Commanding officer conniving with or extorting upon
    sutler ....   "   "
33. Refusal of commanding officer to deliver soldier ame-
    nable to civil justice..   "   "
36. Embezzlement of public stores ....   "   "
39. Embezzlement of public money.................................   "   "
45. Drunk on duty..   "   "
77. Breach of arrest ....   "   "
83. Conduct unbecoming officer or gentleman..................   "   "
48. Non-commissioned officer conniving at evasion of duty
    on the part of privates ....Reduction.
55. Forcing a safeguard in foreign parts.......................Death.

In all other military misdemeanors courts-martial award punishment at discretion. Different expressions are used indiscriminately, such as *dismissal, cashiered, discharged,* and *displaced.* This discretion is not beneficial to an army, as we may illustrate in one case. It may happen that the penalty of death is visited upon a deserter in one army corps, while he, perhaps, would have received a milder punishment if placed before the court of another. The punishment for military crimes should be much more clearly defined than that for civil ones, because in the former case they are generally much swifter and more terrible. For each crime defined there should be a specific punishment, and no other; and *nothing* should be left to the *discretion* of any tribunal, but *everything* to the *grace* of the commanding officer and the President.

To show the singular vagueness of some of our Articles of War, I beg leave to point out only a few instances:

1. In article 5 an officer may abuse the governor or legislature of any state, provided he only takes care not to be quartered in said state. Consequently it would be a military crime to abuse the Governor of Georgia in Georgia, but not on the Potomac.

2. In article 6 I am only prohibited from showing contempt or disrespect to my *commanding* officer, but nothing is said of the many *superior* officers I may have.

3. In article 9 the more comprehensive term of *superior* officer is used. It is no doubt intended in article 6 to include *superior* officers, but as that article now reads I can not punish a man for showing disrespect to any superior officer who is not his commanding officer also.

4. Article 16 displaces an officer for bribery, and *utterly disables him to hold office under the Confederate States' government.* In other articles, where dishonesty is alike punished, the additional penalty of loss of right to hold office is not mentioned, although the misdemeanors are quite as serious as in article 16.

5. In article 21 it is not certain whether officers are to be punished alike with soldiers for absence without leave. In other articles officers alone are punished for misdemeanors which can as well be committed by soldiers.

6. Article 24 arrests any officer who uses reproachful or provoking speeches, or gestures, etc., but it neither orders by whom nor how long they shall be punished.

7. Article 43. The same is the case when soldiers fail to retire to quarters after retreat.

8. Article 49 gives death penalty at discretion to any *officer* who occasions false alarms, but soldiers, who are much more apt to commit that offence, are not mentioned.

Finally, *general court-martial* and *court-martial* are used *ad libitum.* But I

11*

have cited sufficient instances in which the articles could be twisted, if we look at the letter of the law alone.

To show the different kinds of punishments in the Army of France, I attach the following, extracted from the French Military Code :

Death—Abandonment of post by cowardice, obliteration of horse marks, carrying arms against France, assassination, chief of mob, author of mob, seditious clamors, plot to desert, false countersign compromising security, correspondence with the enemy without permission, stripping a body with mutilation or assassination, desertion to the enemy, to foreign armies either off or on duty, after grace, while on sentry, aggravated disobedience in face of the enemy, kidnapping, poisoning, spiking of gun without order, spy, false testimony causing death, incendiarism, insulting sentinel with assault, insult by subordinate with assault, intelligence with enemy, cowardice of sentry in presence of enemy, machinations with enemy, menaces with assault, pillage with arms in hand, concealing spy, reception of deserter at camp after retreat, formal refusal to march against the enemy, revelation of parole to enemy, treason, drummer or bugler without order passing advanced posts, rape followed by death, assault on commanding officer by inferior.

Iron, 12 years—Rape of girl less than 14 years of age.

Iron, 10 years—Stripping of living body, disobedience of a troop (meaning small body of soldiers), violation of countersign, theft at one's host.

Iron, 8 years—Abandonment of post to pillage, fabrication of false signs, forged leave of absence, usage of another's leave, stripping of dead body without order, theft of clothing, double enrolment, deterioration of flour, falsification of march route, signing with false name, insult by subordinate by word or gesture, persistent marauding, menaces of the subordinate, substitution of name upon leave.

Iron, 6 years—Author or accomplice permitting escape of prisoners of war.

Iron, 3 years—Cowardice in abandoning arms in action, false certificate of illness, fraud, cheating in weighing rations, loss of countersign near enemy. sleeping of sentinel near enemy.

Ball and chain, 10 years—Desertion with effects of comrade, desertion to foreign parts, desertion after amnesty, desertion from public labor.

Ball and chain, 5 years—Desertion of substitute.

Labor on public works—Acquitted deserter who does not return arms, effects, or horses he carried off when deserting (term according to offence); desertion into the interior, 3 years; desertion from the army or from a fortress, 5 years; desertion from service, or over the rampart, 5 years; desertion with effects of state or corps, 5 years; sale of arms, clothing, or equipment, 2 to 5 years; desertion with side-arms, 1 year, hard labor.

Forced labor—Murder; theft of arms and ammunition belonging to the state, moneys, or whatever effects may belong to the soldiers or the state (according to offence, the term may be reduced from 3 to 5 years).

Prison—Absence at long roll, 1 month—second time, 6 months; sale of small equipments, 2 months to 1 year; attempt at liberty or safety, 6 months; changing countersign near enemy without giving notice, 6 months ; formal disobedience to superior, 1 year; abuse of armament, equipment, or clothing, 6 months to 2 years; fraud at the house of a citizen when quartered there, 3 months ; fraud with menaces, 6 months; pawning effects or arms, 2 months to 1 year; insulting sentinel, 2, 4, and 6 years; receipt of pawned effects, 2 months to 1 year; assault toward subordinate, 1 year; uniform worn without rank; refusal to employ force; false certificate.

Civil degradation—Abuse of the power of the armed force; arbitrary detention.

Imprisonment, seclusion, or forced labor, according to circumstances—Connivance at proposals made by condemned prisoners.

The maximum of the punishment—Desertion with horse or fire-arms.

This short extract will give some idea of the difference of the French code of punishments and that of our army; and the impression can not be suppressed that our Articles of War and Regulations need revision, so as to adapt them to the various exigencies of the momentous and singular struggle in which we are engaged. I do not know the history of the Articles of War and Regulations of the Confederate Army, but I believe I have sufficiently indicated that they are not all they should be, and that the men who framed them could not have had anything like the very extended knowledge of military affairs required at this day, the army both physically and morally considered. The very defect which exists in our Articles of War—that of leaving the determination of punishments to the discretion of so many variously constituted minds, and thus depriving us of the *uniformity* of punishment so essential to the discipline of an army—has been carefully guarded against in the service of France, and almost *nothing*, certainly nothing *important*, is there left to the discretion of the tribunals. Practical experience, no doubt, has shown the great defects in our regulations ; and I sincerely believe that no board of army officers could be more usefully employed to themselves and the service than by revising our military laws and regulations.

The remarks of a general officer of our service, who has had much experience in the field, and of the adjutant-general of a division, whose practical experience entitle them to great consideration, when speaking to them of our travelling courts as instituted last year, almost entirely agreed with the opinion of the Duke of Ragusa. In the corps in which the general officer above-mentioned commanded a division, this permanent court had done little good, but much mischief. Composed of civilians who know nothing of military matters and of the requirements of discipline, they had punished very serious military crimes with great leniency—others not at all—making the commanding officer perfectly powerless, because he could only either approve the sentence or disapprove it ; and then, in the latter case, the delinquent received no punishment whatever—thus making the enforcement of discipline almost a farce. He objected upon these grounds, to their continuance, and hoped for their speedy dissolution.

It is, therefore, a matter yet to be determined by experience, whether our present system of courts-martial will prove beneficial, or the contrary. Because we are just trying it, there is undoubtedly some difference of opinion among military men on that point. But as the principle is a wrong one, I believe there is no doubt that we will, sooner or later, be compelled to abandon it, as the French were before us.

It seems to me that we are already tending toward the accomplishment of the maxim that a soldier can be justly tried by soldiers alone. Lawyers are very well in civil courts, but upon military tribunals they are out of place. One of our great difficulties to establish a proper discipline has been because we had too many lawyers in the army. Suppose a West Point man were to attempt to deliver his opinion from the bench to a crowd of educated lawyers. We soldiers might think it very well, but there would be no end to the merriment of the legal gentlemen.

PART THIRD.

DIVERS OPERATIONS OF WAR.

CHAPTER I.

EMPLOYMENT OF THE DIFFERENT ARMS.

Their organization should be separate—Their instruction uniform—Combination
of arms as regards their employment—Results of a good mixture—Creation of
esprit de corps—Importance of similarity—Roman legion—An expression of
Vegetius—Middle ages—Frederic—First trial—Marshal de Broglio—The Republic—Formation in brigades—Their constitution—Drawbacks of this system—
It is abandoned—Division constant unit—Army upon a small scale—Napoleon
separates the cavalry from the divisions—Inconveniences—Accessory and principal arm—When and how—Limit of numbers as regards cavalry reserves—
Excess of numbers more embarrassing than useful—Armies of mean strength—
Grand armies; an echelon more needed—Why—The requirements to secure
facility in the exercise of command—Organization of *corps d'armée*—Its composition—Movable reserves—General who has eighty thousand men and the
general with but ten thousand—Their particular rôle—Napoleon at Lützen—
Creation of grades in accordance with the commands—Necessity of an intermediate grade between lieutenant-general and marshal of France—Of self-love.

Note.—The Battle of Grochow.

The troops of different arms should be organized separately, so as to
receive a uniform instruction which may be proper for them, and to
assume the spirit which is suitable in their case.

This principle has been disregarded at different times in the formation of legions, and the inconveniences have been perceived. The
officers who command these corps, knowing that arm best in which
they first have served, give it always the preference, and look upon it
with partiality. In artillery it is absolutely impossible to provide for
the wants of instruction, because the necessary establishments would
have to be infinitely multiplied, such as mock fortifications, schools,
and batteries of different kinds. The artillery should be even united
entirely in a single garrison, if it were possible, so as to receive the
same instruction. The government should, then, devote more money
for this object, because a larger number of individuals would partici-
pate in it. I proposed it when I was chief of the French artillery;

but considerations of administration and economy, strengthened by local interests, prevented the adoption of this change.

But if it is necessary to separate the arms in times of peace, to better develop their special instruction, they must be combined during war.

It is by uniting them with intelligence and skill that the best results are obtained ; they sustain themselves reciprocally, and combine their efforts at the right time. By leaving the same corps together which were united under the same general during several campaigns, an "*esprit de corps*" is created, which is followed by a useful similarity. Troops then will be possessed of all the valor of which they are susceptible. The legion of the Romans is the first example of this combination, which, assuredly, has powerfully contributed to their triumphs. "A god," says Vegetius, "inspired them with the thought."

In the Middle Ages, and the periods following up to our days, the greatest generals have had no idea of imitating it, and Frederic never thought of it. A trial was made in the French army, at the end of the Seven Years' War, under Marshal de Broglio, and this general has had the glory of making this profound thought practical. But the custom was not introduced until the commencement of the Republican wars, and in it is to be found the greatest of revolutions in the art of war in our days.

The infantry, formerly organized in brigades, was under the orders, when formed, of two or three generals, of whom one commanded the centre and the others the wings. The cavalry was likewise divided and placed under the wings, and the subordinate commands were distributed for the day of battle. All generals resided ordinarily at headquarters, charged, in their turn, with the conduct of detachments. A general of an army, wishing to confide a temporary command or some expedition to a general more capable, or who inspired him with more confidence, was obliged to wait until the order of the list called him to duty, and he had to postpone the operation, or order fictitious detachments to employ those who preceded him. When the detachments returned the troops separated, and the brigades were assigned their destination according to the pleasure of the general staff. The question arises how, with such a system, a considerable army could have been moved, formed, and made to fight. Then entire days were formerly necessary merely to place troops in the order of battle. The slightest movement often led to confusion, and the manœuvring artillery leaving the park for the battle-field, and placed in battery, sometimes already the day before, did not come up until after the action.

This barbarous and absurd system has been changed in our first wars ; and soon the armies of all Europe adopted, according to our example, the new organization, which makes troops movable and always ready for battle.   A general now has the means to make with facility such combinations with which the circumstances and his own genius inspire him.

In an army the constant unit, which should never vary, but the strength of which may be more or less, is the division.   It is ordinarily composed of two brigades, each of two regiments, and sometimes three; further, of two batteries of artillery and a corps of mounted troops of some seven to eight hundred horses.   It has an administration complete; it is an army upon a small scale.   Thus it can operate with its own means; it is able to act, march, subsist, and fight separately, or can easily take the post for which it is designed upon the day of battle.

The French army was thus organized during the first and immortal campaigns of Italy, and some years afterward.   Later, Napoleon having formed *corps d'armée*, he took the cavalry away from the divisions, and contented himself to apply to the army corps the principles of the legion.   But in army corps the cavalry is too far away from the divisions; it is not under the hands of the infantry generals engaged, and can not, in many circumstances, profit at the proper time from the disorders which originate with the enemy.   I will, further on, speak of matters connected with *corps d'armée*, and of the circumstances which have authorized and even necessitated their formation.

The division is, then, the unit in an army, the first element through which the three arms are united in an intimate manner; but with it the wants of an army are not satisfied.

Each arm, after having been accessory, should become in its turn the principal one, because there are circumstances when a particular effect must be produced.   Thus, cavalry reserves are indispensable, be it to engage masses of cavalry, be it to be precipitated upon badly-sustained corps of infantry, or to cover infantry in disorder, and to surround batteries, etc.

This cavalry should rely upon and be sustained by a corps of artillery, which belongs to and acts with it in concert, according to circumstances, in the different results which are contemplated.   Cavalry is here the principal and artillery the accessory arm.   But when, during battle, the turn of the latter comes, and the reserve artillery is employed to produce some great effect in a given moment and upon a fixed point, it suddenly becomes the principal arm; it crushes the enemy with its fire; then comes the infantry, which completes the disorder; the cavalry intervening, achieves the destruction and assures the victory.

I do not enter upon those details which would point out the circumstances under which artillery is charged to play an exclusive part; but I have said already enough to lead to the conclusion that each arm should be, in its turn, accessory and principal; and if the artillery is

, ordered to act upon an isolated point, that the infantry and cavalry forces, destined to protect and secure it, should be subordinate to it in all their movements.

But cavalry reserves, however important they may be, should not exceed a certain strength upon any given point; above certain limits the most skilful general will not be able to handle them; and then too many horses united can not be subsisted.   I put the strength with which a good management is possible at six thousand horses; with this number anything which can reasonably be undertaken with cavalry upon the battle-field can succeed.

Napoleon organized, in his last campaigns, corps of cavalry composed of three divisions, numbering at least twelve thousand horses.   This idea was monstrous, and without any useful application upon a field of battle; it became the cause of immense losses, without any fighting— these large corps having never served for anything else but to present an extraordinary spectacle, fit to astonish the sight.

The organization of armies should, then, establish as a principle the formation of divisions and reserves of each arm.   I speak of an army of medium strength, because in large armies another echelon is necessary to serve as an element of order and action.   We come, then, to the formation of troops into army corps—that is, fixed and intermediate commands should be established between the supreme chief and the generals commanding divisions.

An army of one hundred thousand men, composed of from ten to twelve divisions, would be difficult to handle if it was not organized into corps d'armée, because confusion would soon arise from the too large number of independent units, manœuvring at liberty, according to the general direction of the supreme chief.   The want of some formation of aggregations of divisions has, then, been promptly felt, to simplify the dispositions of the chief command, and two, three, or four have been united.   Thus an army, composed as I have just indicated, is divided into four fractions; the general-in-chief can move them with facility; he has under hand four corps, of which three form his line of battle, and the fourth his reserve.

In all the different degrees of military rank, a chief is placed in relation with a small number of immediate subordinates, who facilitate the exercise of his command.

The corps d'armée being small armies, should have an organization in conformity to the principles which I have established, and be composed:

1. Of three divisions, combined of the three arms.

2. Of a cavalry reserve, sustained by horse-artillery.

3. Of an artillery reserve.

The reserves, designed to act anywhere, should be of great mobility, and for artillery, which must often take post at great distances, horse-artillery will be employed.

Thus ordinary artillery, which, by its new organization, is very mobile, would do the service with infantry divisions, and the horse-artillery would be exclusively attached to cavalry and the reserves.

The organization of which I have just given a picture is in accordance with existing armies; it results from the nature of arms and the present manner of making war, and the object of the fractions of the army is to facilitate the exercise of the command. Commands, again, are of different kinds, and change character according to the number of soldiers.

A general who combats with ten thousand men should be in the midst of his troops, and often exposed to the fire of small-arms.

A general commands thirty thousand men; he orders the movement of his troops and reserves, and if he is habitually, with the exception of extreme cases, out of the range of musketry, he should be constantly within that of cannon, and remain within the limit of the space where the balls yet fall.

A general directs eighty or one hundred thousand men; he determines the plan, gives the orders before the battle, opens the movement, and awaits the events in a central position. During the action he becomes a kind of personified providence; he makes dispositions for unforeseen cases, and provides remedies in case of great accidents. He should expose himself before the battle, so as to see for himself, and judge with precision the actual state of things; these duties fulfilled, he gives his orders, and lets every one perform the part assigned to him. If matters progress favorably, he has nothing further to do; if accidents happen, he should guard against them by those combinations in his power; if matters progress very badly, and some great catastrophe is to be feared, he must place himself at the head of the last troops and throw them upon the enemy, and his presence, in this critical moment, gives an impulse to them, and the moral effect produced doubles their valor.

It was thus Napoleon commanded. His operations having been nearly always crowned with success, and the armies he commanded being very numerous, he has but seldom been exposed to any imminent danger. But at Lützen, when matters were culminating in a crisis, and the nature of the army, composed of young soldiers, was augmenting the danger, he rallied the troops himself before Kaya, and led them to the charge under a murderous fire.

From what I have said heretofore it will be apparent which principles were the basis for the creation of the different grades. They have been made in proportion with the actual commands, so that a chief, having a social position separate from those of his subordinates, that position is always superior even when he is off duty.

France is the only country where, greatly to the prejudice of the service, there has not been created an intermediate grade between those of lieutenant-general and marshal for the command of corps d'armée. The dignity of marshal requires a command-in-chief, and the sad experience has been made that, when several marshals were united in the same army and under the command of one of them, great misfortunes were nearly always the result, through the little harmony and subordination which reigned between them. An emperor or captain-general was needed to command an army, the greater portions of which were under the orders of marshals. Corps, it is true, were sometimes under the orders of lieutenant-generals, who received the temporary title of general-in-chief, and a commission for the command. I have even to add, that those who had once commanded them were never again called upon to take a simple division. But the grade being always the same, it is grievous to establish such distinctions freely and voluntarily.

Since authority, necessary everywhere, is still more so among troops, from the command of an army to that of a company—the chief who disappears must be immediately replaced—it has been necessary to establish, as a fundamental principle, the right of seniority in command. But the accidental exercise of this right, brought about by fortuitous events of war, is quite different (every one feels the necessity of this disposition) from the delegation of authority with the same grade, according to the will of a sovereign, when he is the master of the choice.

Self-esteem suffers by being obliged to obey an equal, especially if he is, besides, a junior; and self-esteem, the cause of so much good and evil, exercises, in the profession of arms, an immense influence, because it is the very life of it.

An army composed of men without self-esteem is worth nothing; the French are such good soldiers because they are impressed with it; and through it we find soldiers coming from large cities, where self-esteem is more active, but who are less strong and robust, so much more valorous than those who come from the country.

NOTE.

The Battle of Grochow.—The Russian army, under the command of Marshal Diebitsch, the famed hero of the Balkan, had, on the 5th of February, 1831, upon four different points, passed the frontiers of the contracted portion of old Sarmatia, now called the Kingdom of Poland, for the subjugation of a people who, like us, were fighting for the right of self-government.

His plan of operation was ably conceived. Pushing forward the main portion of his immense host upon the military road which runs due west from the ancient Polish province of Grodno to Warsaw, and touches the town of *Siedlice*, where rested the Polish right, he was, while entertaining the centre and left of the Polish line, to make a vigorous effort against and outflank the right of the Poles, and thus both

12

cut their communications with their capital and interpose his forces between them and their powerful fortress of Zamosk, on the River Wieprz—thus hoping to deal out, in considerably less than ninety days, to the Polish rebellion the death-blow, as our enemies no less confidently expected, at the beginning of this struggle, to do the same with us.

However, there were in his front soldiers commanded by generals who, in the school of Napoleon, had mastered grand operations. Against him he found pitted men such as Chlopicki and Skrzynecki. Their plan of operations was speedily form-ed. Orders were despatched to the different Polish corps who confronted the vast masses of Diebitsch to fall back in the direction of Warsaw, to cover the capital—a movement which, in its general features, strongly resembled that of our forces in the spring of 1862, for the protection of Richmond.

The Russians, deeming the Poles discomfited by their show of superior strength, now began a general forward movement from all points, in pursuance of their pro-gramme of turning the Polish right. While their centre marched upon the high-road and seized Siedlice, some fifty-five miles east of Warsaw, their columns of the right wing debouched upon Warsaw from the north-east. Upon the Polish right, at the village of Stoczek, was posted a corps of observation, commanded by General Dwernicki. This small corps, consisting of barely five thousand men, was, early on the morning of the 14th of February, assaulted by the Russian General Kreutz with fifteen thousand men and twenty-four pieces of artillery. After a sanguinary strug-gle the night saw the Russian column in utter disorder, driven back upon their main body. The first action of the war had resulted gloriously for the Poles; and the Russians—what, with superstitious belief, they thought more disastrous even than the loss of the battle—had left in the hands of the Poles an image of the Holy Mother, which had cheered them under the fire of cannon, and was borne in the very centre of their columns while marching to the combat.

The Polish right wing was now, in conformity with the general plan, more con-tracted toward the centre. But, before this was executed, another heavy action at Boimie resulted, on the following day (15th), in the defeat of the Russians by the small corps under command of General Zymirski.

Two days later, on the 17th, the Polish right and centre were simultaneously at-tacked. At Minsk, the Polish right, the Russians were again discomfited. The centre column, far in advance of the Polish right and left, was under command of General Skrzynecki. On the same day, at Dobre, while slowly retreating, it was closely followed by thirty thousand Russians and sixty pieces of cannon, under Diebitsch and the Grand-Duke Constantine. With a masterly *coup d'œil* Skrzynecki arrested his march near the Village of Dobre, and calmly awaited the debouching of the Russian masses behind a strong position, lined in front with ponds and marshes, with but a single passage across, which could be completely reached by the fire of his twelve pieces of heavy calibre. Urged to deeds of heroism by their illustrious leaders, the Russians, for an entire day, vainly attempted to force the passage, and six thousand in killed and prisoners were by them left upon the field. Night set in upon another day glorious for the Poles.

On the 18th the entire Polish line was retrograding with exemplary order, and on the 19th the army was from Warsaw but forty miles; and on the same day the spirit-ed combat of Swierza took place, where General Dwernicki beat Prince Würtem-berg, who had passed the Vistula at Pulawa, and was approaching Warsaw.

On the same day was fought the battle of *Wavre*, by Sir A. Alison styled the Bat-tle of Grochow; and on the 24th and 25th were fought the Battles of Bialolenka (24th), and of Grochow (25th), both likewise miscalled, by the same writer, the Battle of Praga. His account of all these actions is so mixed up that it is barely possible to discern the true occurrences.

The Battles of Wavre and of Bialolenka were nearly fought upon the same ground as the Battle of Grochow. From the 14th to the evening preceding the great battle ten days had passed, during which some ten sanguinary battles had been fought, in which the Poles, though retreating, had been uniformly successful. The heroism and endurance of their small army during these memorable ten days is above all praise. In that short space of time the whole Russian army had been engaged by a force not above one-sixth its strength, and thirty thousand of the invaders had been either killed, wounded, or taken prisoners.

For a like series of sanguinary actions, against such an overmatched force, we look in vain in history. By it is demonstrated what a brave people may do in the defence of its homes and the liberty of its country. Only a cause just and grand could be contested for with like devotion. In the middle of winter, in a climate infinitely more rigorous than ours, in addition to forced and harassing marches, the days were spent alternately in fighting and marching, with but little rest for all; for such were the requirements of the hour that, continually, while one-third were resting upon the snow-covered ground, two-thirds had to be kept under arms to guard their comrades' slumber. For the straggler in the rear, had there been any, there would have been no quarter. Numberless hordes of Cossacks were hovering around, day and night, ready to pounce upon the isolated victim as the hawk does upon its prey. A continuous retreat, necessities of war, so disastrous in nearly all its annals, appeared to have for this brave race nothing dispiriting. Their *morale* was not shaken an iota; on the contrary, they seemed to draw inspiration from the hourly increasing proximity to their capital, where, they well knew, tender hearts were awaiting them, to cheer them on with noble devotion in the great battle which was destined to be fought under the very walls of the ancient city.

The army of the Poles was now concentrated, with remarkable success, for not a single column of theirs had been cut off or defeated, on the west side of the Village of Grochow, one Polish mile, or some seven English miles, from the City of Warsaw, and it occupied a line which had been chosen with admirable genius. The right of the Polish line rested upon the impenetrable marshes of the Vistula, called the Marshes of Goclaw. The left was posted upon the slight elevation which commands the Village of Kawenzyn, near the Vistula. From right to left there stretched, for about three miles, an unbroken plain, composed mostly of cultivated fields, without any other obstacles save those presented by the furrows which divide the fields of the different proprietors—thus forming a magnificent field for all the manœuvres of the three arms. Between the centre and the left the Polish line was perpendicularly traversed by the high-road already mentioned, which leads from the province of Grodno, through Siedlice, upon Warsaw, and for the possession of which road, opening the way to the city, the battle was being fought.

There was, however, upon this line, and between the Polish centre and left, a position which, in Polish history, was destined to become as famed as is the little "wood of birches" in the Battle of Hochkirch in Prussian annals—namely, a forest of elder trees, situated in front of the Polish army, and for the possession of which incredible efforts were made, during the struggle, by vast Russian hosts.

The Russian position was upon a front parallel to that of the Polish forces, concealed by the dense woodlands which surround the Village of Grochow, and which the Russians occupied in force. The road from Siedlice to Warsaw, having passed over a country thickly wooded, after leaving the Village of Grochow debouches upon the vast plain, stretching for seven miles away toward the capital; and from the belt of woodlands, which concealed the Russian line of battle, the invaders beheld, on the cold, frosty morning of the 25th of February, 1831, the magnificent prospect of the towers and domes of the rich and populous capital of the country, in front the Vistula winding through the expanse, and which they had come to destroy—confidently trusting that it would, by the mere force of overpowering numbers, be their conquered prize before the evening sun had set in.

The Russian army consisted of eight corps of combatants, and three in reserve. Their left wing was between the Village of Wavre and the marshes of Goclaw, and composed of four divisions of infantry, forty-seven thousand strong, four divisions of cavalry, fifteen thousand seven hundred, with one hundred and twenty pieces of cannon. The centre, which rested opposite the forest of elders, consisted also of four divisions of infantry, fifty-seven thousand men, three of cavalry, ten thousand five hundred strong, and had one hundred and eight pieces of cannon. The right wing, opposite the Village of Kawenzyn, was composed of three and a half divisions of infantry of thirty-one thousand men, four of cavalry, fifteen thousand seven hundred and fifty men, with fifty-two pieces of cannon. Upon the borders of the great forest, opposite the forest of elders, was placed the reserve, commanded by Grand-Duke Constantine, in infantry and cavalry twenty thousand strong, with thirty-two cannon—making a grand total of *one hundred and ninety-six thousand infantry and cavalry, with three hundred and twelve pieces of artillery.*

Against this enormous host the Poles could muster, after great exertions, but a vastly inferior force, numbering not more than forty-three thousand four hundred infantry and cavalry, with ninety-six guns. General Szembek, commanding the right wing, had seven thousand infantry, with twenty-four guns, and occupied the space between the high-road and the marshes of the Vistula. The centre, composed of Skrzynecki's and Zimirski's divisions, fifteen thousand strong, with sixty guns, occupied the space between the forest of elders to the high-road; the left wing occupied the Village of Kawenzyn with a force of six thousand five hundred men and twelve guns, under the command of General Krakowiecki. The entire cavalry, nine thousand five hundred strong, commanded by Generals Uminski, Lubinski, Skarzynski, and Jankowski, were deployed in rear of the infantry and artillery, ready to precipitate themselves upon the Russian columns whenever an opportunity should offer. Besides these, a small reserve of four battalions and eight squadrons, in all about five thousand four hundred men, under the command of General Pac, were posted upon both sides of the Warsaw high-road, a little to the rear of General Szembek.

Thus, in sullen silence, the opposing hosts remained during the night of the 24th, awaiting the day which was to decide upon the fortunes of a country, and which to thousands was to be the last one which should dawn upon them. The night was unusually serene and clear. Thousands of watch-fires, around which the weary Poles were reposing, illumined the horizon, while along the dark line of the forest which enclosed the Russian hosts everything would have seemed quiet as the night but for the dense volumes of smoke ascending from behind the curtain. Far in the distance myriads of lights were to be seen, showing the extent of the Polish capital, which, like beacons, shone all night—for but few were there in the ancient city who, wrapt in slumber, were forgetful of the threatening dangers.

With the break of day the armies were awaiting, in serried columns, the beginning of the struggle. The first rays of light had scarce dispelled the darkness when, upon the Polish left, in the direction of Kawenzyn, debouched from the forest the Russian right, with a force as large as the entire army of the Poles. Fifty pieces of artillery preceded the columns, on the wings of which hung clouds of cavalry in threatening masses. A tremendous cannonade was directed upon the village, which, soon in flames, was wrapt in clouds of smoke. To this overpowering force there were opposed seven battalions and twelve guns. With unflinching resolution Generals Krakowiecki and Malachowski made the most gallant efforts to keep their ground. At the head of their columns, on foot, they repeatedly charged the advancing battalions of the enemy, while the twelve pieces of artillery, skilfully served, tore whole streets through the enemy's masses—unmindful of the concentrated fire of the Russian artillery—directing every discharge into the dense advancing hosts. For five long hours this gallant band withstood the successive shocks, till at last, weary of unsuccessful efforts, the enemy's fire slackened and soon died away. Thus the great attack upon the Polish left had been repulsed.

Marshal Diebitsch, during the whole of this attack, had firmly expected that the Polish left would be forced. He had, therefore, directed the whole masses of the Russian right upon the Village of Kawenzyn, expecting to force the Poles to weaken their centre in order to sustain their left. But they well knew that no succor could be expected from any quarter in the defence of the positions assigned to them respectively, and thus they had formed the resolution to conquer or to die upon the ground.

Up to ten o'clock in the morning the centre and right of the Polish line had remained unattacked. But when the efforts against their left became apparently useless, Diebitsch resolved upon a great demonstration against the Polish right. Two hundred pieces of cannon, as if by magic, began to vomit their missiles of death against the Polish line. The earth seemed to tremble under foot. Covered by this tremendous fire, the Russians now began to debouch from the forest, and in one moment the plain of Wavre was covered with their columns. Looking over that plain, between the forest of elders and the Vistula, the eye saw nothing but one undivided mass of troops in motion; not even the different divisions could be distinguished from each other. Still under cover of the guns, the Russians steadily advanced. But a great catastrophe awaited them. In an incredibly short space of time the entire cavalry of the Poles had been collected; issuing through the openings left between the columns of the infantry, they threw themselves upon the Russians, and, with one grand charge, swept them from the field.

This brilliant success inspired the Poles with the greatest ardor. When the brave horsemen returned from the charge cries of defiance rent the air from the entire Polish line, striking terror into the enemy's ranks.

A lull now succeeded in the unequal contest. Rapidly the Russians concentrated one hundred and twenty guns against the forest of elders, held by Generals Skrzynecki's and Zimirski's divisions, composed of the very flower of the Polish army. Among the devoted soldiers holding this now for ever memorable forest was the celebrated Fourth regiment of infantry, which, on the day inaugurating the revolution, had, to a man, taken the solemn oath administered by their brave colonel, Boguslawski, never to fire a single shot, but always to attack with the bayonet, until their country should be free—a pledge carried out under their succeeding colonel, Borcenzki, and especially in the fierce struggle of Grochow, and so faithfully that, at the end of the revolution, but ten men remained.

One hundred thousand men now assaulted the forest of elders, and after a fierce struggle the Poles were driven from their position. But the artillery, with a bravery never surpassed, dashed within two hundred yards of the Russian columns, and a tremendous discharge of grape, double-shotted, shattered the advance columns of the confidently marching host. The Polish infantry, but fourteen battalions strong, with a fierce charge of the bayonet, repulsed the Russians, and the forest was re-

12*

taken. But in an instant fresh regiments of infantry were hurried against the blazing forest, while the Russian cuirassiers attacked both flanks. While they were endeavoring to break through the line the Polish cavalry, ever on the alert, defeated the iron-clad regiments, and drove them back in utter confusion. The Polish infantry a second time was driven from the forest. Again the artillery shattered the Russians, and again, with one charge, they were repulsed by the infantry, and the forest retaken a second time. The other portions of the line, during these fierce attacks, were not unmolested. On the right and left the combat raged furiously, without, however, making any visible impression upon the stout Polish infantry.

Toward two o'clock in the afternoon, while the combat was extending along all points of the line, the enemy made one grand last effort against the forest of elders, when the second division, terribly reduced, began to give way, and the Russian columns at once poured into the interval which had thus been made. Destruction appeared inevitable; the crisis of the battle seemed to have come; and already the wavering preceding a great catastrophe was to be seen in the Polish centre, when the battle was restored by one of the most daring artillery charges on record. The five Polish batteries of Adamski, Maslowski, Hildebrand, Bielak, and Pientka advanced, like cavalry, to the charge, throwing themselves into the interval with undaunted resolution, and, approaching close to the rapidly advancing Russian columns, opened a fire of grape which spread destruction and disorder in their ranks. The Russians halted in astonishment at so daring a charge; and while they were yet at bay, the batteries, with the celerity of thought, had taken shelter in rear of the now confidently advancing second division, which, for the ninth and last time, had now retaken the blood-stained ground of the forest.

For four hours this terrible massacre had lasted. Under their heroic leaders, Skrzynecki, Zimirski, Boguslawski, Czyzewski, and Rohland, the Polish centre executed deeds of daring such as have never been surpassed. Opposed, a living mass, to the concentrated fire of one hundred and twenty guns, they never wavered for a moment under a storm of missiles more terrible than had been hurled by four hundred French guns, nineteen years before, against the great redoubt of the Moskowa. Changes of front, the attack in columns, and in echelon, the concentration of forces upon the Russian points which were wavering, reserving their fire until close to the enemy and not wasting a single shot, were executed with an activity, order, and coolness unparalleled. Only by such conduct could the tremendous attack of the Russians have been withstood for four hours by fifteen thousand men, who, at the ninth attack, had been reduced to less than ten thousand.

Like the infantry and artillery, the Polish cavalry had performed prodigies. Besides protecting the advancing artillery, the different charges which they executed with such bravery, they were manœuvred with the utmost skill by their generals, and continually called upon to shift from point to point in order to fill the voids occasioned by the inferiority of the Polish forces, so as always to present to the enemy an unbroken line.

By such manœuvres of the three arms combined, and executed with the greatest celerity and determination, in which every commander performed his duty with the greatest devotion, the plans of the enemy were never permitted to be fully developed and executed; and his dazzling force, which at first sight would have been supposed capable of absolutely crushing the small Polish army, was, in effect, but one great unwieldly mass, which only confidently marched to the combat under the protection of an enormous number of guns, the fire of which, during the whole day, upon every portion of the Polish line, had been steadily maintained, whether necessary or not, with the utmost regularity.

The battle had now raged for nearly twelve hours, and the loss of life on both sides had been immense. The three grand attacks of the Russians, it is true, had been unsuccessful, but at the cost of weakening the extended Polish line to such a degree as to make it exceedingly dangerous to await a new attack upon the same front. The Polish generals therefore determined to withdraw their forces to the rear upon the same line, the centre of which had been occupied during the struggle by the small reserve corps of General Pac, near an obelisk of iron, immediately upon the high-road, or some half a mile from the former line of battle. The object was to gain a more commanding position; to draw the enemy upon the open plain; to concentrate the forces still more, and to place them upon two lines—the second to be composed of the whole of the second and a part of the third division, which, having withstood the main attack in the centre, were nearly exhausted. Furthermore, it was expected that the enemy would be led into the error to suppose the manœuvre a retrograde movement, forced upon the Poles by their losses, and that they felt themselves too weak to continue the defence of the forest.

To execute this manœuvre, and to enable the second division to retire from the forest of elders without being molested, the artillery was left with some twenty squadrons to protect the retrograde march. The artillery and cavalry were ordered to evacuate their positions gradually, and the former to take post in the centre under the protection of the whole of the cavalry, which, formed in echelon, were prepared for a general attack. The change was as admirably executed as it was conceived. The enemy had no suspicion of its object, but, presuming it to be a flight, at once undertook to profit by it. It was then that Marshal Diebitsch, sure of victory, saw himself already at Warsaw, and, in the exultation of the moment, exclaimed: *"Cho-rozo! tiepier ja dumaiu po skonczanii etoy krwawoy srazenia, ja wsostianji budu wziat w Belwerderskom dwarce."* "Well, then, it appears that, after this bloody day, I shall take tea in the Belvidere palace."

It was about three o'clock when the second division began to retire by echelons. To hasten the execution of this movement General Chlopicki ordered that the columns, retiring in succession, on reaching a considerable distance from the enemy should quicken their pace as they proceeded, in order to form the second line as soon as possible, and to give space for the operations of the artillery and cavalry. It was at this moment that General Zimirski, who had lost several horses under him and had just mounted a fresh one to superintend this movement, was struck by a twelve-pound ball, which carried off his arm, making a terrible wound in his shoulder, of which he died in a few hours. The melancholy loss of this general was most deeply felt by the whole army, and particularly by his own division, but it did not interfere with the success of the movement. The brave General Czyzewski immediately took command of the division, greatly supported by Generals Rohland and Zaluski.

As soon as the last columns of the Poles quitted the forest of elders the Russian troops began to debouch from it, and the Polish artillery commenced a terrible fire. The brave Colonel Pientka, who was still in front, checked the advance of the columns. Seated, with the most perfect *sang-froid*, upon a disabled piece of artillery, he directed an unremitting fire from his battery. The artillery and cavalry, after having protected the retrograde movement of the centre, still held their ground to enable the wings also to retire undisturbed.

The whole Polish army was now in full march for the new line of battle, and the second division had already begun to form the second line, and had commenced firing by battalions. At this moment Marshal Diebitsch, with admirable celerity, had massed a heavy force of cavalry between the Village of Kawenzyn and the forest of elders to penetrate and annihilate the retiring Polish columns. Forty squadrons, at their head five regiments of cuirassiers, issued from the borders of the forest upon the plain, in one solid body—ten thousand admirable horsemen coming to the charge.

In their way was nothing but the battery of Colonel Pientka, supported by a single regiment of lancers. Pientka, with great coolness, greeted them with one tremendous discharge of grape; then limbering up, he quitted a post, at full gallop, which he had held for five hours, to save himself from being cut off.

Animated by this rapid retreat, the Russian squadrons now came onward upon a trot, and in line perpendicular to a battery of rockets which had been stationed in the second Polish line, between the second and third divisions. But Skrzynecki's division had already formed in squares to receive the onset. Nearer and nearer they came toward the wall of fire of the brave second division, when suddenly the battery of rockets opened fire. The horses of the cuirassiers, maddened by the shower of flakes of fire and the hissing noise of the rockets, stopped, reared, turned, partly fled and partly dashed through the Polish lines. Throwing the rear columns of the heavy mass into inextricable confusion, the squares now opened fire, and from the plain were seen dashing onward the renowned Polish lancers, completing the utter rout of the, five minutes before, apparently invincible host. So nearly complete, in fact, was their destruction, that of a regiment of cuirassiers which was at the head of the attacking force, called the regiment of Albert, bearing the designation of "Invincible" upon their gorgeous helmets, not a man escaped. A portion of this cavalry had, however, penetrated, and was dashing onward toward Praga, where, from the walls of the fortress, they were received by a murderous fire of artillery and utterly annihilated, upon their retreat, by the Polish cavalry.

The whole Polish line had now, as if by magic, formed into squares. But as soon as the cavalry column was seen to retreat, overthrowing in their course the Russian artillery and infantry which had been ordered up to support the grand attack, and carrying utter consternation into the Russian ranks, the squares deployed. General Chlopicki, the Polish commander, while the change of line was being executed, had been severely wounded in the leg by a grenade. It was necessary to appoint, upon the field of battle, a new commander-in-chief, and the Polish generals, with singular unanimity, called Skrzynecki to that responsible position, although he was junior to many of them.

Under his orders the whole Polish line pressed onward upon the now retreating and utterly discomfited Russians. A cavalry force of ten thousand fresh men would probably have completed their destruction. But we have seen the immense task which the Polish cavalry had already so ably discharged. Worn-out, they could not attempt a prolonged pursuit.

The Russians left upon the field of battle thirty thousand men. The Poles had sustained a loss of twelve thousand in killed and wounded, or one-third of their army—a loss equal to that sustained by the Confederate arms upon the bloody field of Shiloh, with only this difference : that the majority of the wounds received by the Russians at Grochow was from the bayonet of the Poles, and that of the latter mainly from the artillery and musketry-fire of the Russians.

Such was, in its main features, the hard-contested Battle of Grochow. Warsaw had been temporarily saved. Diebitsch was disgraced; and months elapsed before the Russians again assumed the offensive.

This battle has been sketched as an illustration of what the three arms of the service may accomplish when skilfully combined. There is, besides, no instance in modern history where such an inferiority of forces defeated a like immense host. This achievement will never fail to fill us with admiration for a people who can thus fight for liberty, and who, after a despotism of over thirty years has kept them in chains, have again risen, and are now sustaining themselves for over three years, against the enormous powers of the Russian Czar. The germ of liberty never dies; and if we, with advantages much greater than those of the Poles, can not achieve our independence, we do not deserve to have it. But we most assuredly will achieve it, if, mindful of the great lessons which history teaches, we can rise above the slimy level of extortion and the greed of gain, shake off the apathy which now disgraces those at· home, and come back to the glorious days of 1861, when all was fervor and patriotic devotion. If we do so, success is not doubtful for a minute; if not, what mortal man can tell how long our brave soldiers in the field will retain that spirit which hitherto has made them the barrier between us and utter degradation?

CHAPTER II.

OFFENSIVE AND DEFENSIVE WARS.

The applications of the principle easier in the latter than in the former—Particular characteristics of each—Necessity of studying contemporaneous wars—Why—They instruct better—Wars of Frederic—Resources of his genius—The campaigns of the Revolution—Memoirs of Gouvion Saint-Cyr—Operations of the Archduke Charles in 1796—Campaigns of Italy in 1796 and 1797—First example of strategic operations upon a vast scale—Immense results with mediocre means—In one year, an audacious and skilful offensive and a model in defensive operations—Unexampled victories—Immortal epoch—Summary of achievements—Should be written with commentaries—Campaigns of 1805, 1806, and 1809—Spanish wars—Neglect of principles—Awakening in 1814—One against ten—Thirty-five thousand men of remnants—Champ-Aubert, Vauchamps, Gué-à-Trême, Paris—Song of the Swan.

Notes.—1. The Battle of Lützen. 2. Marshal Mortier.

I have already stated, and I repeat, that the movements of war, be they offensive or defensive, should always be founded upon a calculation as to both time and distance. But the applications of this principle are easier in defensive than in offensive wars.

In the latter the combinations are vaster, the conditions more variable, and the elements of calculation more uncertain. At every moment a general may be forced to change his programme, to abandon an attack in order to defend himself, and to escape from great dangers. A more extended genius is then necessary, to be always ready to vary one's projects, and to execute new combinations.

In defensive warfare the theatre is more contracted; the operations are conducted upon sections of country well known, and the nature of which can be exactly appreciated. The number of combinations being less, they are more easily carried out and provided against. In offensive war the genius must supply experience and divine the nature of the country destined for operations; the points of support which are counted upon vary, and sometimes disappear altogether. In defensive war the field of operations has been well prepared and studied; the pivots of operation are fixed, and everything can be calculated with precision. A superior genius is then more necessary for an offensive war, while an extensive knowledge of the profession of war, the talent to choose well the points of support, great foresight, and indefatigable activity, may be sufficient for the requirements of the defensive war.

This mode of warfare is, however, far from being easy—since, properly speaking, a general is only reduced to assume the defensive when it becomes apparent that his disposable means are inferior to those of the enemy. Besides, in modern wars, with such an equality in arms, instruction, and experience, numbers are of great importance. The difference which exist between such and such an army, and such and such a campaign, depends more particularly upon the *morale* of the troops, and its appreciation is not prescribed by the rules of the profession, but must be sought in that sublime part of the art which treats of the knowledge of the human heart, the movements of which are so rapid and mysterious.

After having laid down the principle upon which the movements of armies depend, it can only be developed by examples. Instruction lies in the study of the most memorable campaigns. Dogmatical teachings are founded upon facts. They can be chosen from successes as well as reverses—by dividing the recital of each event into that part which treats of the combinations, and the other which depended upon hazard.

The study of the events of our epoch should be preferred; the examples will be better comprehended, because the events are better known. Besides, on account of the progress which the art of war has experienced, and the actual and ever-increasing mobility of our armies, those achievements have become easy which formerly were impracticable.

The former wars which are still full of useful instruction are those of Frederic II. It is true that the examples of those times are no longer applicable to our days, because everything is changed; but that great captain's campaigns should especially be studied in regard to their moral aspect.' When, after having been beaten at Hochkirch, with the loss of some two hundred pieces of cannon, Frederic is seen to retire upon two single points only, upon the Spree, and there takes a menacing position toward the enemy, the explanation of such a mystery is vainly asked for in our days.

In reflecting upon the weakness of Frederic's resources, we furthermore inquire how, in the presence of such numerous enemies and during so many years, he could have maintained and recruited his armies. Truly, we hardly know which to admire most, whether the glory of his victories, or his talents for discovering new resources and preserving his forces.

The long wars of our epoch, the great events presented to our contemplation, and the careful examination of all attending circumstances, the military student must equally observe in our armies and those of the enemy.

The first campaigns of the Revolution present nothing, both with us and with our adversaries, which is not susceptible to severe criticism ; the proof of this may be found in reading the first volume of the Memoirs of Marshal Gouvion St. Cyr, which, in this respect, are of the greatest interest.

The operations of Archduke Charles in 1796, when opposing the French Armies of the Sambre and Meuse, and of the Rhine, are the first example of operations systematically combined upon a vast scale. The work of this prince can not be studied too much; his principles are there established by an exposition of his operations and the motives which dictated them. While all grand principles of war may be deduced from them, their application is, at the same time, found in the facts narrated.

But the campaigns requiring most reflection are those of the French army in Italy in 1796 and 1797. They all unite exactness of calculations, correctness of movements, and a profound knowledge of men and things.

Never was war more admirably and perfectly conducted. The most sublime part of the art of war has been illustrated through it. With mediocre means immense results have been obtained.

This war, which lasted hardly one year, presents models of all kinds : an offensive skilfully and audaciously conducted; a defensive where inferior forces have constantly repulsed superior forces, but often managing upon the field of battle to have a superiority of numbers; a war which, by the skill of direction and the vigor of execution, has led to a series of unexampled victories. It was an immortal epoch, the prodigious achievements of which have surpassed everything either before or after; because, in a series of combats so long continued, and executed in the midst of so many divers movements, it is impossible to discover a single fault, or a single neglect of the true principles of the art.

At the moment of the opening of the campaign the French army, scarcely thirty thousand men strong, and in want of everything, had not even completed its preparations when it was forced to commence

active operations—the enemy approaching from Genoa to cover that
place. The hostile army was attacked, composed, it is true, of troops
of two different nations, but more than fifty thousand men strong.
The Austrians were beaten, pursued, and soon consisted of but a single
division. The French army then threw itself upon the Piedmontese
army; complete and rapid successes augmented the confusion and the
discouragement of the Allies, and the King of Sardinia made peace.

By a precipitate march the passage of the Po was seized, which
river the French army could not cross in strong force on account of
the want of material. An energetic action cleared the line of the River
Adda of the enemy. Milan opened its gates. Soon after, an insurrection convulsed a whole province; the insurrection was quelled. The
army, which had scarcely slackened its march for a moment, passed
the Mincio in strong force, arrived upon the Adige, and took a defensive position which covered the conquests made in less than fifty days.
Hostile armies were successively formed, and expended upon us their
useless efforts. Mantua fell; we marched upon Vienna, and the peace
was concluded.

Nothing would be more useful for the instruction of officers who devote themselves to the study of grand warfare and to military concep-
tions of a higher order, than to write this memorable campaign, with
the details and documents belonging to it. Commentaries should be
joined to it, which, explaining the reasons of the movements, would
show the character of the campaign and its results. The fine campaign of 1805, so well conducted and so remarkable in its conclusion,
but, it is also true, favored by immense and almost incredible faults of
our adversaries; that of 1806, which completed it; finally, that of
1809—may well be objects of special study and instructive commentaries, because this great epoch of Napoleon's life can not be too much
admired.

But with silence we are compelled to pass over the Spanish wars
and those following that time, or at least we can only speak of their
faults, and show that fortune abandoned Napoleon on the day when he
became unfaithful to the true principles of the art which hitherto he
had always respected. Then the accumulation of men and means was
useless. From the date of this epoch of sad memory, if the Battles of
Lützen[1] and Bautzen are excepted, Napoleon is not recognized in any
one of his campaigns.

A kind of awakening, however, came later. The great captain was
himself again in 1814; but the spirit of the people and his soldiers
only fought for him; he had no longer an army; we could hardly
place one man against ten. The forces of which Napoleon could dispose, in his movements between the Seine and the Marne, never exceeded thirty-five thousand men of shattered remains. My corps, which

alone had the honor of combating at Champ-Aubert, Vauchamps, Montmirail, and the second affair of Gué-à-Trême, had never as many as four thousand men, the remains of fifty-two different battalions. At Paris, sustained by the Duke of Treviso[2], our united forces only amounted to fourteen thousand men, and the enemy had fifty-three thousand men engaged, and thirteen thousand were disabled. It was the song of the swan.

NOTES.

1. The Battle of Lutzen.—The opening of the campaign of Napoleon in 1813, after the great emperor's misfortunes in Russia, was characterized by his want of cavalry, and by the many conscripts, young and inexperienced soldiers, of which his army consisted. The veterans of Jena, Austerlitz, and Wagram lay buried beneath the snows of the vast Eastern plains. Opposed to him he found the coalition of Prussia, a state whose politics have at all times been signalized by wavering and treachery—and Russia, eager to avenge the invasion of her territories. They both controlled vastly superior means to those of Napoleon, and an extraordinarily powerful body of cavalry was ready to take advantage of any of the defeats which the French should meet.

It was at this period that Napoleon proclaimed to his legions the maxim, forced upon him by circumstances, and wrung from him against his conviction, to assure his young conscripts: " *Q'une bonne infanterie soutenue par de l'artillerie doit savoir se suffire.*" ("Let it be demonstrated that a good infantry, sustained by artillery, is amply sufficient.")

The great military road leading from France into Germany, upon which Napoleon advanced, crosses into Germany at the confluence of the River Main with the Rhine, protected by the strong fortress, Mayence. From thence it leads in a north-eastern direction through the very centre of Germany to Weissenfels, Lützen, and to the City of Leipzig, a strategic position of great importance; from there it leads to Dresden, capital of the Kingdom of Saxony, where it crosses the Elbe, and from which city three important roads go to Berlin, North-eastern Prussia, and the heart of Bohemia. Napoleon's purpose was first to humiliate the Prussians as at Jena in 1806, and then to debouch upon the Russians. This was, however, in measure prevented by the exceedingly vigorous movements of the Allies, who, crossing the Elbe at several points, occupied Leipzig and threatened Napoleon's communication with France, before the latter had scarcely half concentrated his forces. The central point upon which the French forces moved was Leipzig. Viceroy Eugene, whose gallant attitude in keeping the Allies in check had permitted Napoleon to make his proper dispositions, was at Merseburg, a town some eleven miles north-west from Leipzig. As soon as Napoleon had crossed the Saale the Allies moved forward with great promptness to prevent his march upon Leipzig, and resolved to give battle in the plains of Lützen, a small town already immortalized by the heroic death of Gustavus Adolphus of Sweden in 1632. It is a very remarkable position. The road, after having ascended the defile of Poserna, runs along a plateau exceedingly favorable for combat, and densely dotted with little villages. *The strategical combination of the Allies was to refuse their right, to make a feint attack upon the centre, but to throw their whole weight upon the French right, thus to entirely change the position of the French, and to cut them off from the great road to France, upon which they just were advancing.* Napoleon was expecting no serious resistance until he should have occupied Leipzig and was advancing upon Dresden, when both armies came suddenly, on the 2d of May, into collision, both in open columns and under march.

South of Lützen, and to the westward, there runs a little streamlet, with steep banks on both sides, called the Flossgraben. This the Allies had crossed in force in four heavy columns. In front of this streamlet lie four small villages—Gross-Goerschen, Klein-Goerschen, Rahno, and Kaya—which was the French right, forming an irregular quadrangle, and of which Kaya is nearest to Lützen and to the high-road to Leipzig, and of which the first, Gross-Goerschen, was only occupied by one division of the French under General Souham. The Allies overwhelmed Souham, took the village, and with shouts of triumph carried Klein-Goerschen and Rahno likewise. The situation became critical; but Napoleon had already mastered the enemy's combinations, and directed all of his forces to his right. The troops halted upon the high-road and wheeled into line. Marmont hastened across the fields. Eugene heard the firing from afar, and retraced his steps.

The brave Ney now supported Souham with three divisions, and by a splendid charge regained the lost villages.

The Prussians, returning to the attack under Wiltgenstein, threw the French into confusion. They not only carried the villages again, but drove the French beyond Kaya, the key of the French position.

The French now abandoned their whole line of battle, taking a new position some six hundred yards in rear. It was six o'clock in the evening; the battle had already lasted eight hours. The most alarming despatches reached the emperor, who, upon his left, awaited the issue of the struggle on the right. He at once set out in person, and reached the position where his troops appeared to be in the greatest disorder. But, as at Marengo his presence reanimated the heroes of Italy, so here the young conscripts cried " *Vive l'empereur!*" when they saw him, and formed for a renewed struggle. Kaya having now become the decisive point of the battle, Napoleon led his troops to the attack himself to retake it. A most desperate conflict ensued. The reinforcements had come up, and with seventy thousand men the Allies were pressed, and the brave Ney again drove them beyond Kaya.

But in the meantime the Russian artillery, sustained by the infantry of the reserve, had taken an advantageous position upon the French left, and were making a great impression. The Village of Kaya was retaken a third time, and the French army repulsed from all the villages again.

The decisive moment now came. Sixty guns of the Imperial Guard, under the immortal Drouot, sustained by sixteen battalions of the Young Guard, advanced to the final issue, as the Old Guard often had done before the disastrous Russian campaign, followed by the entire cavalry of the reserve. Kaya was regained, and the successful advance of Eugene upon the rear of the forces which assaulted the French upon the left decided the struggle in favor of the arms of France.

Here Napoleon once more showed his genius in all its splendor. Surprised at Gross-Goerschen in the morning at ten o'clock, with but Ney's corps at hand, and his troops scattered upon the high-road at a distance of thirty miles, he was in imminent danger of having his army pierced. He saw that the struggle was to be for the possession of Kaya, in spite of the feint attacks of the enemy upon his left, and that he contrived to restore the battle after the heavy repulse from Kaya at six in the evening, entitles him to the highest glory.

His inferiority in cavalry, six thousand against thirty thousand of the Allies, prevented him from pursuing them, and to reap any fruits from this and his subsequent great victory at Bautzen.

It was a campaign in which his talents were taxed to the utmost; but nothing could replace his want of cavalry, and the fatal issue of Leipzig may be traced to this cause. Without its vigorous and decisive concurrence, after an impression has once been made upon the enemy, nothing but barren victories may be expected.

But then the cavalry must be highly disciplined, well drilled in firing, and especially in the handling of the sword.

We have the most splendid material for cavalry in the world. Neither the Cossack nor the Hungarian Honved are superior to the Southern horseman. The Ranger of the Confederate States surpasses them all.

The foreign levies of the United States do not know how to ride, and will never learn it. How many Confederate colonels of cavalry are there who are swordsmen? These remarks may offer to them some food for meditation and *imitation*.

2. Marshal Mortier, Duc de Treviso.—The career of this officer was of the most brilliant nature. He was a captain in 1791, and rose, during the wars of revolutionary France, to the rank of general of division. He was employed with the Armies of the North, the Sambre and Meuse, and the Rhine. In 1801 he was in sole command of the seventeenth division of the Army of Hanover. Created a marshal of the empire in May, 1804, he served with great distinction with the Grand Army, then in Spain, and again with the Grand Army of France. After the departure of Napoleon for the Island of Elba he joined the Bourbons, and held many appointments of importance under the Restoration. In 1831 he was Grand Chancellor of the Legion of Honor. In 1834-5 Minister of War and President of Council. Napoleon created him Duc de Treviso. He was Chevalier of the Order of St. Esprit, and bore several foreign (Austrian and Portuguese) orders. He died in July, 1835.

CHAPTER III.

MARCHES AND ENCAMPMENTS[1].

Marches within reach of the enemy—Precautions—Composition of advance guards —Out of the enemy's reach—Utility of light troops—Wooded and broken countries—Scouts on the flanks—Thrown out in fan-shape—Camps—Suitable spots— Natural obstacles—To camp this side—Why—Surprises—Two modes of camping —Deployed and massed troops—Which preferable—Way of execution—Consequences of disregard of rules—Haynau, 1813—The division Maison surprised— Fine revenge upon the Prussians at Vauchamps, Hohenlinden, Falkenheim— Marches in the enemy's presence—This is tactics—Vigilance and foresight of the chief—Discipline and quick manœuvres on the part of the army—Parallel march of the French and English armies in 1812—Historical details—Flank movements —French army marches like a regiment—The two generals hesitate to give battle—They keep within a space of five leagues—The only instance of a march of this kind.

Notes.—1. March Regulations of Doctor Jackson. 2. The Battle of Hohenlinden.

Marches within reach of the enemy can not be executed with too much precaution, nor the encampments be selected with too much prudence. Every one knows how the former are executed.   Still, marches are modified by the nature of the country, the composition of the advanced guards, and by the respective position of the arms of which they are composed.

The object being to obtain news from the enemy, and to have knowledge of his arrival upon his approach, it is useful to gain this information at the greatest possible distance from the enemy, without, however, compromising the security of one's detachments.   The vanguard of an army, if it be not in the presence of the enemy, should at least march at a considerable distance from the mass of the troops, and that of a division should be at a march of several hours' duration ahead.

Light troops must be employed with intelligence, and neither should they be spared—for in this service particularly consists their usefulness; if they permit the army to be surprised, the officer who commands them fails in his duty, and he can allege no sufficient excuse.   In broken and wooded countries, especially, the precautions must be doubled. Scouts thrown out upon the flanks should be sustained by detachments upon which they can fall for support; the detachments should, besides, be strong enough to defend, when necessary, those defiles for some time, which would enable the enemy to turn the army.

By combining the march of troops sent out to reconnoitre the enemy, so as to be always formed in the shape of a fan, every surprise is guarded against; and they are never exposed, because their point of retreat is always upon the line of operation of the army.

Upon marches, encampments are established to repose troops and to satisfy their wants, but never to engage the enemy.   An encampment is placed, by preference, upon the banks of a small stream, near to a village, to supply the soldiers with water, and to bring within their reach the resources usually found among an agglomerated population.   But whatever be the importance of these considerations, security should likewise be considered, and the means must not be neglected to provide against unforeseen attacks and surprises.   I do not allude here to the guards; they must always cover and surround a camp; they are of the greatest necessity, be it only for the police.

If there be any obstacle, the encampment should be placed upon this side and never beyond it—at least, for the greater portion of the troops. It would, undoubtedly, be advantageous to have passed the defile upon commencing the day's march, and to debouch more easily; but this advantage is more than counterbalanced by the security of repose.   If there be no obstacle, or if the obstacle may be easily turned, a surprise is to be feared; a numerous cavalry may suddenly appear, as if rising from the ground; then security must be even sought in the selection of the encampment.

There are two ways of camping: the troops are deployed upon the color-line, or formed *en masse* by battalion. This last disposition is much to be preferred, and offers many advantages.

It is executed in the following manner:

A division is placed upon two lines, and each battalion is formed in columns by division; it is, furthermore, separated in two demi-battalions in columns by platoons. The interval separating the two fractions is equal to the front of a division, and forms a perpendicular street to the line of battle of the encampment.

The tents or barracks are established upon the right and left, and their front is placed so as to open upon the street, either directly or by a tranversal lane. Whenever the battalion takes arms, each soldier falls into his platoon, which is formed in the camp street, almost at the same time with the battalion, ready to march. If impetuosity leads a body of cavalry to precipitate itself upon the camp, it will find all the troops massed, and, so to speak, entrenched in the midst of their tents and barracks.

The violation of the above rules led, on the 29th of May, 1813, near Haynau, in Silesia, to a deplorable event. The division Maison, which, having marched during the entire day, and taken position without sufficiently reconnoitring the same, was surprised; twenty-two Prussian squadrons, ambuscaded in the neighboring wood, debouched suddenly at the moment when the division was about establishing itself in camp; the consequence was, its being almost entirely destroyed, without having been able to offer any resistance.

Upon another occasion the same negligence on the part of the Prussians gave to us a fine revenge, and obtained for us an easy victory. After the combat of Champ-Aubert (10th of February, 1814), where my *corps d'armeé* alone destroyed and took prisoners almost the whole Russian corps of Olsufieff, the emperor ordered me to repair to Etoges to cover the army in that quarter, while he was to march upon Montmirail, occupied by the corps of Sacken. Sacken, beaten, retired upon Château-Thierry, where he passed the Marne to arrest Napoleon's pursuit, who had followed him. During this time Blücher in person had advanced with Kleist's corps, and had marched upon Etoges; on the 13th he took his measures to force me to evacuate this advantageous post. After having feigned that I wished to defend it, in order to retard Blücher's march, I commenced to retreat; the enemy kept near to me, but followed with great circumspection, and until the evening we engaged, only feebly, our light troops. I took position upon the borders of the woods of Fromentière, and the enemy encamped within two cannon-shots of myself. I had communicated the arrival of Blücher to Napoleon, and, having made him acquainted with the movements I had just executed, I was assured of his prompt return. On the 14th, at four

13*

o'clock in the morning, I began to march upon Montmirail, and despatched an officer to obtain information from the emperor. He was approaching, and sent me word that I might attack the enemy whenever I would find it convenient, and that he was within supporting distance.

There is, in front of the Village of Vauchamps, toward Paris, an advantageous and easily to be defended position ; it is upon the declivity of the plateau which borders the valley in which Vauchamps is built. Upon the right of this position, a wood in front gives the means to take in reverse any body which, inconsiderately advancing, would fail to secure it at first. I occupied this wood in a cautious manner; I deployed upon the hill-side, placed my cannon in battery, and awaited the enemy.

The corps of Kleist, whose strength was four times greater than my own, believed they had nothing to apprehend, and marched with exceeding confidence, his troops disposed in columns and touching each other, without any distance between them, and even without having any flankers ; finding the village unoccupied he traverses it, but, assailed by a murderous artillery and musketry fire, attacked at the same time in the front and upon the flanks, he was put in confusion, retired through the village in the greatest disorder, and our cavalry falling upon him, four thousand prisoners fell into our hands. From this moment the enemy retreated from morning till night, without any regular formation, and this favorable action for us was followed for him by a new catastrophe.

The victory of Hohenlinden[2], the glory and results of which were so great, is an event of like nature. The centre column of the Austrian army which followed the high-road, and with which a large portion of the artillery was joined, marching in lateral columns to facilitate the transportation, outstripped the other columns, and it moved without having sufficiently reconnoitred its line of march, on account of the confidence which the combat of the preceding day had given the troops, and from the belief that the French army was beaten and retreating. It suddenly encountered the latter in the depth of the forest. Attacked with vigor before having been able to make the necessary dispositions for resistance, soon taken in flank, this immense column of material was surrounded, and the battle gained.

Nothing is more delicate and merits more attention than traversing, with a numerous artillery, a very wooded country in presence of the enemy. Whatever be the urgency of quickly uniting under like circumstances, too many precautions to guard against a surprise can not be taken, because the consequences of the least negligence are nearly always deplorable.

On the twenty-ninth of August, 1813, after the Battle of Dresden, I was charged with the pursuit of the hostile army, the greater part of which was retreating upon the road to Attenberg. After having beaten the corps which covered the movement of concentration at Possendorf and Dippoldiswalde, I was to continue my march the following morning upon Falkenheim. Arrived at the Village of Frauendorf, I learned that the enemy occupied, with a strong vanguard, a good position at Falkenheim. Before bringing on the engagement in the forest which I was compelled to traverse, and which was occupied by some light troops, I had it thoroughly ransacked and cleared by some three or four thousand infantry, extended upon a very large front. After having freed it from the enemy, I occupied the borders of the forest with my vanguard, and awaited the entire junction of my corps. I then debouched with all my means; the enemy was overthrown and driven from his position in a moment, and left behind almost his entire artillery.

There are also marches executed in presence of the enemy, with an army entirely united, thoroughly formed, and ready to combat, the object of which is to force an enemy to quit a position he occupies. These marches belong to tactical movements, but nothing merits greater attention and requires more precautions.

To execute a movement of this kind very disciplined and well drilled troops are necessary, commanded by active and intelligent generals, and by a chief endowed with great foresight.

The Army of Portugal, in 1812, under my command, executed such a march with success.

The French and English armies were encamped upon both banks of the Duero; the first was inferior to the other by six thousand infantry and four thousand cavalry. Despite the inferiority of forces, I had been obliged to assume the offensive. , I had been instructed, through my official correspondence, that any important succor would not be given to me; and, on the other hand, the English army, already so superior, was enabled within a few days to receive powerful reinforcements from Estramadura by the bridge of Alcantara, while the Army of Galicia, which blockaded Astorga, had just become disposable and operated upon my rear in consequence of the reduction of that city, which, on account of the scarcity of provisions, was upon the point of opening its gates. I concluded that, to change the state of things, it was necessary for me to assume the offensive, but with prudence; so to manœuvre as to force the enemy to retreat, and not to engage him until it became necessary. The passage of the Duero was then resolved upon and executed.

The French army, fully united, encountered the next day two English divisions at Tordesillas de la Orden, who retired in haste; they were hotly pursued and would probably been destroyed, being quite isolated, had the French cavalry been less inferior to that of the enemy.

The two armies found themselves, upon the evening of this pursuit, facing each other, and separated by the Guarena, a marshy streamlet.

On the twentieth of July the French army, formed throughout in order of battle broken into platoons, made a manœuvre by the flank to the right, to ascend the streamlet; arrived at a ford, known in advance and promptly taken advantage of, the head was thrown upon the left bank, seized at once a plateau which extended indefinitely in a direction menacing the retreat of the enemy, and debouched upon it under the protection of a very heavy battery which covered its movements.

The Duke of Wellington at first believed to be able to oppose this offensive march, but it was executed with so much spirit and unison that he soon renounced the idea of attacking us.* He contented himself in following the French army in the direction of a plateau parallel to that which we held.

The two armies continued their march, separated by a narrow valley, always in readiness to receive battle; several hundred cannon-shots were exchanged, by taking advantage of the more or less favorable positions arising from the sinuosities of the plateau, and both generals were willing to receive battle, but neither to offer it. They arrived thus, after a march of five leagues, in the respective positions which they wished to occupy—the French army upon the heights of Aldea-Rubia, the English army upon those of Saint-Cristoval.

This remarkable march is, to my knowledge, the only one of this kind which has been executed in our times. But it may occur again, in a war where the forces are balanced, and when the generals are only willing to engage with advantages already assured, or when certain and very favorable circumstances take place.

NOTES.

1. **March Regulations of Dr. Jackson.**—A knapsack crammed with necessaries, so as to load a foot-soldier like a pack-horse, oppresses by its weight, consequently consumes a part of the power which is intended for, and which ought to be reserved for, military exertion. Superfluity of baggage is a common error in the British service; and the usual manner of disposing of it for carriage is not, moreover, well contrived. A full knapsack rolls upon the back like a billet of wood, and shoulder-straps gall the skin, if the whole weight of the pack bear upon the shoulder. To remedy the rolling of the back and the galling of the shoulders, the shoulder-straps are joined by a belt across the breast. The remedy is worse than the evil it is intended to remedy; and it is worse for this reason, that few persons are aware of the mischief which it occasions. The pressure of a cross-belt confines the free motion of the chest, and impedes respiration. Whatever impedes free respiration increases the heat of the body beyond the just temperature. It is thus that a person who joins the shoulder-straps of his pack by a belt across the breast is oppressed with heat, and pants for breath, frequently without adverting to the cause which occasions the increase of heat and oppression. On the contrary,

---

* The Duke of Wellington told me afterward that the French army marched at that moment like one regiment. This was his own expression.—*Note of Author.*

where the pack is supported wholly by the shoulder-straps, though the shoulders may be galled the respiration is free, and the body is less liable to be overheated.

As soldiers are supposed to be arranged in companies according to powers of exertion, and as there must necessarily be some variety in the effective power of companies, it is obvious to common sense that the least effective companies ought to be placed in front, the movement being there least embarrassed. The rate of the slow pace is three miles per hour—the rate of the exerted pace, four. These paces are to be changed at stated intervals only, time and distance being measured exactly by an officer, who leads at a justly-regulated step. If this be not done with care, a precise effect can not be expected in combined movement; and hence it happens that by the neglect or by the transgression of this fundamental rule of order, the military purpose is defeated, or less completely executed than it might be.

Various contingencies arise, in the course of a march, which oblige individuals to leave the ranks. The act of leaving the ranks is unmilitary in appearance, and reprehensible irregularities not unfrequently follow the practice of it. In order, therefore, to remove all shadow of pretext for the occurrence of such necessity, it will be proper that a general halt be made for five minutes at the end of the first hour, so that every one may, during the interval, adjust those personal concerns which require adjustment. The march of the first hour is supposed to be performed at the slow pace; that of the second, at the accelerated. The column halts for fifteen minutes or more at the end of the second hour; and, during the halt, the individuals are supposed to recline, or assume a horizontal position—for it is only in the recumbent position that the limbs experience the full benefit of rest. When fifteen minutes have expired, the march is resumed at the slow pace. When the hour is completed, the column halts five minutes for purposes of personal adjustment, and, at a given signal, resumes its course at the accelerated pace. In this manner a journey of fourteen miles is performed in the space of four hours and twenty-five minutes, including the time allowed for halting; and if the march be conducted in the manner proposed, no person, it is presumed, who is fit to be admitted into the military ranks, will fail in performing it. A distance of fourteen miles is a common day's march for troops on ordinary service. Circumstances sometimes occur which require that the distance be lengthened, even that it be doubled. The exertion will not, it is believed, bear hard upon well formed troops, if due care be taken in adjusting the primary arrangement, and due consideration employed in directing the subsequent steps of the march. For example, it is understood that a halt for the space of one hour takes place after the performance of the first part of the allotted march, and that the shoes, socks, and trowsers, or breeches, and leggins, be then taken off; the feet, legs, and thighs washed, or bathed in cold water, if the nature of the halting ground supply water in sufficient quantity for that purpose. If water be deficient, the lower extremities may be rubbed with a wet towel, and exposed to the cool air. Such is a simple expedient only, but it restores vigor and capability of exertion equal to some hours of rest. If hunger or faintness be felt by any one, a crust of bread, with a morsel of cheese, washed down by tea, or vinegar and water—with which every soldier is understood to be provided—is sufficient to remove it. The march is to be resumed at the expiration of an hour; and, with the observance of the rules prescribed, the distance, it is presumed, will be performed with ease in the calculated time, if care has been taken in the primary arrangement to separate the weak and inefficient parts from the sound and effective.

2. The Battle of Hohenlinden.—From Munich, the capital of Bavaria, a high-road leads in an easterly direction into Austria, through the following places: *Hohenlinden*, eighteen miles; *Matenpot*, twenty-four miles; *Haag*, twenty-eight miles; turning north-east, it strikes the Village of *Ampfing*, forty miles; and five miles further the Town of *Muehldorf*, forty-five miles from Munich; beyond it the road crosses the River *Inn*.

Another road leaves Munich, south and parallel to the former, to Salzburg in the Tyrol, passing through *Ebersdorf*, eighteen miles east-south-east, and *Wasserburg*, thirty-one miles east-south-east from Munich, crossing the Inn likewise beyond the latter place.

The *forest of Hohenlinden* spreads out in contiguous masses, forming a natural stockade at from fourteen to seventeen miles in length and four miles in depth, and its *depth* is traversed by these two roads. Hohenlinden and Ebersburg lie upon the entrance to and on the Munich side of the forest; in other words, the forest is the defile and those two villages the points upon which an army, coming from Austria, debouches after having traversed the defile. Between these two roads are some impracticable by-roads. From Muehldorf, as far as Hohenlinden, the country is hilly, intersected by woods and rivulets; from Hohenlinden, as far as Munich, it is but one fine and continuous plain.

Moreau's forces were, on the 1st of December, 1800, in straggling and detached columns, moving from Hohenlinden upon these two roads toward Austria; Lecourbe and Sainte-Suzanne's divisions were detached far in the rear. Moving thus from west to east, Grenier, who commanded the left, while leisurely approaching Ampfing was suddenly attacked by the Austrians, under Archduke John, who were being moved from north to south, and thus thrown perpendicularly upon the French, perfectly united and sixty thousand strong. Grenier was routed at once. In vain Ney, Grandjean, Hardy, and Legrand strove to make a stand; and the decline of the day saw the French army, thoroughly beaten and panic-stricken, returning through the forest toward Hohenlinden.

Archduke John was content, and failed to take advantage of the victory. He gave to the enemy one day, the 2d of December, to reorganize. Moreau, whose genius was up to the emergency, knew that the Austrians would have to traverse the forest to follow up their victory. He took a position in front of Hohenlinden, where the forest begins, and awaited his adversary.

Before daylight on the 3d the Austrians marched in three heavy columns through the forest. The roads were terrible, and it snowed incessantly. The centre, forty thousand strong, with one hundred cannon and five hundred caissons, marched upon the Muehldorf high-road; the right upon the lower road, under Latour, twenty-five thousand strong; the left and the light troops upon cross-roads.

The centre, outstripping the other columns because it marched upon the best road, debouched upon Hohenlinden at nine in the morning, when Grouchy drove it back. The Austrians, however, steadily gained ground; but Grouchy and Grandjean, at the head of fresh battalions, repulsed them once more. Meantime the Austrian right began to debouch, but Ney opposed it most effectually.

The French general, Richepanse, meantime had advanced, early in the morning, between the space occupied by the two high-roads, to take the Austrian centre in the rear—whether by Moreau's orders, or independently, is a disputed point, but, it is most probable, by the orders of Moreau. He arrived within sight of the Village of Matenpot, where the rear of the Austrian centre rested, composed of artillery and the cavalry—both reposing, the latter dismounted—and was about to attack, when he himself was pierced by the Austrian left, under Riesch. Thus brought between two immense bodies, a general of mediocre talent and courage would have permitted the centre to march unmolested, and either turned against the lesser body or sought safety in retreat. But Richepanse engaged both by himself, assailing the cavalry and artillery, and charging the commander of his severed forces to hold the position they then occupied against Riesch. This he did nobly, and no efforts of the latter dislodged him. The success of Richepanse's efforts was immense. The utmost consternation seized the rear of the centre; it fled toward Hohenlinden.

At the moment when Richepanse arrived near Matenpot, Moreau nearly found himself overwhelmed by the Austrians. But when he heard the cannon in the direction of Matenpot he directed Ney and Grouchy to make a last charge, which overthrew the now consternated enemy. The fugitives meeting the onward pressure of the rear columns pursued by Richepanse, a perfect rout ensued, and the day was gained. Grenier, who commanded the French left, and already overpowered by the Austrians, upon the intelligence of the rout of the centre resumed the offensive and drove Latour back into the woods. Moreau's head-quarters were that night at Haag. The Austrians lost over one hundred cannon, three hundred caissons, and eighteen thousand men ; the French nine thousand.

The most important lesson which this battle teaches is *the importance of following up a victory*. We see here that the Austrians were as successful on the 1st of December as a general could wish to be. Moreau and all of. his generals not alone, but the men, were in consternation. It is most likely that if Archduke John had secured the defile of Hohenlinden and pressed on toward Munich, Moreau would have been obliged to cross the frontiers of France. But one day's delay brought upon the Austrian arms the heaviest defeat they had as yet experienced.

The second observation is, the very confident manner with which the Austrians marched through the forest, unconnectedly and without sufficiently reconnoitring their lines of march.

The third observation to be made is, that had the French trusted alone to the strength of their position and contented themselves to hold it, it is most likely that they would have been conquered a second time. Moreau was already exhausted in the centre, and the left, under Grenier, beaten; the Austrians were already concentrating in the plain of Hohenlinden in front of the woods. But Richepanse's turning manœuvre decided the day. It was the most brilliant manœuvre of the campaign. It was a defensive attitude skilfully combined with a vigorously executed offensive movement, which decided the whole campaign. The fact that General Moreau was awaiting with impatience the movement of Richepanse, appears to establish, beyond controversy, that he had ordered it, and so it is asserted by General Jomini. It would be strange, indeed, if an important feature in a plan of battle, such as the turning of an enemy's flank, could be accomplished without at least the cognizance, if not the orders, of the general commanding.

In our war, where direct attacks are so prevalent, Lieutenant-General Jackson has executed several such manœuvres. They will always stamp upon a general the seal of immortality.

CHAPTER IV.

GRAND RECONNOISSANCES, AND PRECAUTIONS THEY REQUIRE.

Great difficulty in the conduct of an army—We must seek the enemy—Necessity of permanent contact—Rôle of cavalry—The curtain must be lifted—Precept—Consequences of its disregard—The Army of Portugal in the Valley of the Tagus—Twofold reconnoissance upon Alméïda and Elbodon—English cavalry flies—Brigade of infantry isolated—Fuente-Guinaldo—The reconnoissance not properly carried out—The English army escapes—Smiles of fortune.

To know the position of the enemy, to be informed in time of the movements which he is executing, to gather sufficient facts from which his projects may be divined, is one of the greatest difficulties which the command of an army presents continually. Nothing should be neglected to arrive at exact information, and the surest means is to be constantly in contact with the enemy by means of light troops, to have frequently small engagements, and to take prisoners, whose answers are generally always *naive* and sincere. More can be ascertained from them than from the most faithful spies; the latter often confound the names of corps and generals, and estimate the strength of the troops upon which they report very inaccurately.

When two armies, through the combinations of war, find themselves suddenly in the presence of, or have remained for a long time at a certain distance from, each other, it is well to be assured more positively of the situation of affairs; and then are made what we call *grand reconnaissances*. These operations demand great prudence, and even particular foresight, especially if it be undecided whether to engage or not, at least in extraordinary and very advantageous circumstances.

It is necessary to employ a large force of cavalry, and, if possible, only light cavalry and horse-artillery, so as to remain more surely master of one's own movements. The question is, how to withdraw the curtain which covers an army, and when a general has sufficiently penetrated to the front to see with his own eyes the situation of the enemy, he has attained his object.

But he should also be in a proper state of sustaining the troops which are engaged, and to gather them, should they have become scattered in a lively attack. He will have within reach a corps of infantry of respectable size, and in the rear of this corps the whole army will be disposed in such a manner as to be enabled to march at once, should circumstances require their taking any part in the action. One moment of hesitation may lead to the neglect of the employment of those sudden circumstances which, seized in the proper manner, would have brought about unforeseen advantages.

I may cite one instance where the disregard of this precept prevented me from achieving an easy victory over the English army in Spain. The recital of faults is, perhaps, more instructive than the account of successes.

In 1811 I occupied the Valley of the Tagus with the Army of Portugal. My mission was to watch over the safety of two strong places which covered the north and the south—Ciudad-Rodrigo and Badajoz—which belonged to the Armies of the South and the North, and made part of their department. Ciudad-Rodrigo being in want of provisions, General Dorsenne, commanding the Army of the North, organized a large train and made preparations to conduct it. He furnished, as an escort, a force of ten thousand infantry and two thousand horse. But the co-operation of the Army of Portugal was necessary to assure his march, the English army being encamped in his proximity. I carried the greater portion of the army beyond the Col de Banos, and I marched in echelon from Tamames until I reached the River Aguéda. I marched upon Rodrigo with fifteen hundred cavalry. General Dorsenne arrived there likewise, and securing his large train of munitions of war in that place, left there a small division of three thousand infantry, commanded by General Thiébauld. The rumor had been circulated that the English were disposed to invest Rodrigo, and that all army stores within reach had been concentrated in the place. To secure them would have been opportune for them, and it became necessary to make a double *reconnaissance* upon the road to Alméida and toward the heights of Elbodon, upon which the outposts of the English army had been established. This *reconnaissance* was to be executed by the cavalry of the Army of Portugal, commanded by General Montbrun.

General Thiébauld received orders to sustain the latter when necessary. The position of Elbodon having been carried in a moment, the English cavalry was put to flight, and a brigade of English infantry found itself isolated. After having bravely sustained several charges, it retreated upon Fuente-Guinaldo. Favored by the difficult nature of the ground, and thanks to the rapidity of its march and its bravery, we were unable to take it. It became important to occupy without delay the Village of Fuente-Guinaldo, it being the point of convergence of many roads, and of strategical importance for the concentration of the army. The division of Thiébauld was therefore called upon; but placed too far away, because the object of its march had only been the defence and security of a convoy, and the field of battle having become singularly distant, owing to the enemy's retreat, it arrived too late, and its extreme weakness forbade its being thrown, at the beginning of the night, upon the entrenchments of Fuente-Guinaldo, toward which place the columns, coming from different sides, were directed. Had eight thousand men been at my disposal I would have been able to act with confidence. Fuente-Guinaldo would have fallen into my

14

power; the light division, placed at Martiago, upon the right bank of the Aguéda, would probably have been taken or destroyed, the English army dispersed, and its corps, being without communication, would have been in the most critical position. Having had time to unite, it hastened to make good its retreat, and the opportunity of an easy and complete success was passed.

I repeat that, when making a *reconnaissance* in force, the troops should always be disposed so as to prevent the enemy from forcing them to a serious engagement; but, at the same time, they should be able either to rally troops should they have been beaten, or to profit from fortuitous and favorable circumstances. Whatever consideration we accord to our adversary, we should never believe him to be infallible; fortune often smiles when least expected, and we should always be prepared to prove that we are not unworthy of its favors.

CHAPTER V.

DETACHMENTS IN PRESENCE OF THE ENEMY—THEIR CHANCES, AND THE DANGERS WHICH ACCOMPANY THEM.

Have for object to profit from an expected victory—Examples of failures not counted upon—Wurmser in 1796, near the Lake of Garda—Alvinzi at Montebaldo and Corona—Detachment upon the Adige—Detachments upon the Po, Tesino, and Adda—Supposition of a retreat upon Genoa—Forty-five thousand men against twenty-two thousand—The Austrian army attacks—Battle of Marengo lost at five, P. M.—Return of Desaix's division, detached upon the Genoa road—Fine victory—But a dangerous example to follow—Rout at the Katzbach in 1813—Conclusions, and precepts.

Note—The Battle of Marengo.

Sometimes a general, too much preoccupied with the hopes of a success, makes beforehand, without having as yet beaten the enemy, such dispositions as will give great results to his victory. To attain this object he divides his forces and sends them in different directions; instead of conquering, he is beaten. His detachments are either surrounded or destroyed, and a campaign opened under good auspices, ends by a series of reverses.

In support of this I can cite several examples:

In 1796 Wurmser opens the campaign in Italy with an army superior to the French; a column turns the latter, and throws itself upon its communications toward Brescia. This column, too feeble to resist the united French army, retires at its approach. Separated from the great er portion of the army by mountains and the Lake of Garda, it remains a stranger to succeeding events, and the French army, posted in the centre of the enemy's position, beats, one after the other, every corps which successively appears in its front.

In the same year, 1796, General Alvinzi debouches from the Tyrol and attacks the French army, occupying the chain of the mountains of Baldo and La Corona. Believing a victory to be certain, he detaches a corps of five thousand men, commanded by Colonel Lusignan, who, after having followed the borders of the Lake of Garda, changes direction, approaches the Adige, and takes a position in rear of the French army and upon its direct line of communication. This corps is held in check by the feeble division of Rey, which rejoined the army and established itself in front of the Austrians. The battle is gained by the French army, and the corps of Lusignan attacked, is routed and almost entirely captured.

In 1800 Napoleon debouches into Italy with an army of sixty thousand men. Having passed the Po and completely turned the Austrian army, he finds himself upon its communications, and is about to seize all the roads by which it could seek to retire*. To attain this object he places part of his forces upon the left bank of the Po, upon the Tesino, while he was compelled to despatch a division upon the Adda and the Oglio to cover himself in that quarter. Then, supposing that the Austrian army, assembled at Alexandria, would most likely retreat upon Genoa, he detaches a division in the direction of Novi, to close that road likewise. But twenty-two thousand men were at his disposal, while the enemy had a force of forty-five thousand men assembled upon the Bormida. The enemy attacks, and the Battle of Marengo takes place; disputed with obstinacy, at five o'clock in the evening it seems as lost, when the division despatched toward Novi arrives. General Desaix, who commanded it, had wisely stopped its march when hearing the cannonade, to await further orders. He retraces his steps, and arrives yet in time to be useful as a reserve force, and the battle is gained, although but twenty-seven thousand men had in all been engaged, and of these twenty-two thousand alone had been obliged to sustain the shock of the battle. Thus our forces, in this instance, amounted to but two-thirds of those of the enemy, and but for one fortunate circumstance they would have numbered but one-half. It was undoubtedly a fine victory, the results of which were immense; but it would be dangerous to take its strategical combinations as a model,

---

* The army which fought at Marengo only consisted of Victor's corps, which was composed of the two small divisions of Gardanne and Chamberlhac, and of Lannes' corps, composed of the divisions of Vatrin and Monier, the division of Boudet, five thousand strong, a very small force of cavalry, and thirty-two pieces of cannon.— *Note of Author.*

since the battle ought to have been lost, on account of the superiority of forces and the means which were brought against us.

If victories under such circumstances are likely, their possibility must not be too much relied upon. An energy, correspondingly increasing with less favorable circumstances, may undoubtedly be shown, but the latter must not be wantonly provoked.

In 1813 the French Army of Silesia, more than eighty thousand strong, concentrated at Goldsberg, commanded by the Marshal-Duke of Tarentum, confronted an army nearly equal, commanded' by Blücher. The Duke of Tarentum advances upon the enemy, supposing him to be massed at Jauer; he detaches, at the moment of his making a motion, the division of Puthod, to march by way of Schoenau upon Jauer, so as to engage the enemy in flank.

But Blücher at the same moment himself takes the offensive; the French army having reconnoitred badly, suddenly encounters the enemy near the Katzbach, and is obliged to accept battle without having united its forces. Bad combinations and a series of unfortunate circumstances lead to confusion. The French army being beaten, is forced to fall back; the division of Puthod loses its communications; thrown back upon the flooded Bober, overwhelmed by numbers, and after having valiantly striven to resist, it is captured entirely.

From the examples above cited, and many others which could be added, the following conclusions are drawn:

1. Nothing is more dangerous than to make a detachment of any importance before the battle has been fought, a victory been achieved, and a decided advantage over the enemy been obtained.

2. The execution of this hazardous combination requires that the army have a sufficient superiority to assure great probabilities of victory, and that concentrated forces be never weakened beyond the strength of those of the enemy.

3. When distant from an enemy who is strong enough to give battle, and when marching toward him, the space of at least one day's march from the enemy's forces must be occupied by the vanguards and light troops, so as to be informed of his movements, and to modify ours correspondingly.

4. Lastly, when it is believed that an isolated detachment can be made with advantage, its direction must be determined and troops be stationed to support it, in such a manner as always to assure its retreat upon the army, and that it in no case be in danger of losing its communications.

NOTE.

The Battle of Marengo.—Early on the 14th of June, 1800, Napoleon was here surprised, while one-half of his forces were detached, and the remainder, about twenty-two thousand men, under Lannes and Victor, were disposed in *oblique order by echelon, left in front, and the right at half a day's march in the rear, in marching order*. Marengo is near the fortress of Alexandria, upon the marshy borders of the Fontanone, in a perfectly level plain. The Austrian general, Melas, with over thirty thousand men and two hundred cannon, fell suddenly upon the French. Orders were immediately sent to Desaix to come up.

The Austrian infantry overthrew Gardanne, the advance guard, in front of Marengo. Victor stemmed the tide for over two hours unsupported, and was at length reinforced by Lannes' vanguard. The Austrians steadily progressed, and Melas, hearing that the French general, Suchet, had arrived at Aqui, a village some twelve miles south of the battle-field, detached a body of twenty-five hundred horsemen to hold him in check. Meantime Victor was obliged to retreat through Marengo upon a position some five miles west and a little to the south, to the Village of San Guiliano. Lannes retreated across the plain, amid a terrible fire, with admirable precision and coolness, in echelons by squares. Although the French horse, under Kellermann and Lampeaux, made most gallant efforts to disconcert the Austrian infantry, the latter progressed steadily onward, and a total rout of the French was already impending.

At eleven o'clock, A. M., Napoleon with his staff, and two hundred grenadiers of the Guard, arrived upon the battle-field, and the sight of him reanimated his soldiers. He brought with him the vanguard of Desaix's division, and at the same time when Napoleon rallied his centre, and at the head of a demi-brigade of Lannes' division, advanced toward the Austrians, Monier, with five battalions of Desaix's vanguard and eight hundred grenadiers of the Guard, fell upon the extreme left of the Austrians at Castel Ceriolo, and carried that village. Retaken by the Austrians in the ensuing struggle, Cara St. Cyr again stormed it, and held it the remainder of the day.

While the French right was thus victorious, the left was thrown into irreparable disorder, and giving way on all points. Napoleon had already resolved to withdraw northward across the fields toward Pavia, and abandon the battle-field. Melas, assured of victory, had given up the command to Zach, to rest at Alexandria.

In this emergency, at four, P. M., Desaix made his appearance at St. Giuliano. "*The battle*," counselled this brave soldier, "*is undoubtedly lost, but we have time to gain another one.*" Victor and Lannes were reformed and massed in front of San Giuliano; Marmont prepared a masked battery of twelve guns, and Desaix advanced at the head of four thousand men to the charge. Zach, unsuspectingly advancing, was received by the discharge of this masked battery, and Desaix debouched from the villages; while advancing toward the staggered enemy, he fell, pierced by a ball in the breast. The enemy, recovering from the surprise, advanced with six thousand Hungarian grenadiers, and the French, in their turn, hesitated and broke.

In this moment Kellermann charged. He describes it thus: "The combat was engaged; Desaix soon drove back the enemy's tirailleurs on their main body; but the sight of that formidable column of six thousand Hungarian grenadiers made our troops halt. I was advancing in line on their flanks, concealed by festoons of vines; a frightful discharge took place; our line wavered, broke, and fled. The Austrians rapidly advanced to follow up their success, in all the disorder and security of victory. I see it; I am in the midst of them; they lay down their arms. The whole did not occupy so much time as it took me to write these six lines," and the French achieved a complete victory. Loss: Imperialists, 7,000 killed and wounded, 3,000 prisoners, 8 standards, 20 pieces of cannon. French, 6,000 killed and wounded, 1,000 prisoners.

14*

This battle illustrates: 1. The importance of concentration of forces. 2. The danger of receiving an enemy when disposed by columns in echelon, because any disorder in front speedily spreads to the rear, and the successive columns, instead of coming up to the aid, will mostly find themselves overwhelmed by retreating forces. 3. The folly of the Austrian general to detach twenty-five hundred of his cavalry to check a movement in his rear, when General Suchet was fully twelve miles distant, and could not have come up for several hours—time enough to beat the French. 4. That a victory was won, through Desaix's counsel, from four, P. M., to sunset. Thus, one single hour, late in the day, decided the fate of Italy.

CHAPTER VI.

BATTLES.

Rules and principles—Variations—Nature of terrain—Strength of position augments the number of troops—Defiles in advance—Formation upon two lines—Not absolute; commands should embrace both lines—Why—Reserve—General disposition—Defensive battles—Choice of position—Rear free and protected—Offensive battles—Strategy and tactics—Dash, intelligence, and skill in manœuvres—Genius of the French for the offensive—Immense difficulties in that warfare as regards administration and maintenance of troops—The defensive the genius of the English—Wellington in Spain—Masséna menaces Portugal—The Duke shelters himself behind the Coa and two fortresses—Retires upon Lisbon, awaiting the disorganization of the French—System decided upon and perseveringly followed —Renewed at Waterloo—The English give a defensive battle—Precepts—Transition from the offensive to the defensive—Example—Offensive war the genius of Napoleon—His campaign in Italy offensive, then defensive, then again offensive —1805—Austerlitz—Jena—1809—Ratisbon—Wagram; a front attack—1812, Battle of the Moskowa—Importance of a flank movement—Direct attacks—Likes to employ great strength—1813, Lützen, defensive—Bautzen—Leipzig—Bad battlefield—The 18th of October, defensive—Battles of Brienne, Craon, Laon, Arcis, Champ-Aubert, Montmirail, Vauchamps, Montereau—What to do at Paris— 14,000 men against 54,000—13,000 of the enemy *hors de combat*—The time of battles—Regulated according to circumstances—With superior forces attack soon in the morning—Why—Napoleon at Waterloo—With equal forces attack in the middle of the day—Why—1796, reverse of Cerea and Due Castelli—Wurmser out of Mantua—Victory of Saint-George—Defensive battles a part of the profession— Frederic II in the Seven Years' War—His campaigns resemble Napoleon's—Every general has his own manner of conducting war—Turenne and Condé—Alexander and Cæsar—Fabius, Hannibal, and Scipio.

NOTES.—1. Confederate corps of reserves. 2. The lines of Torres-Vedras, in Portugal. 3. Wellington's strategy. 4. The Battle of Waterloo.

To treat in detail of all the dispositions which the conduct of a battle demands, is a thing impossible. A thousand unforeseen circumstances may force us to modify them—fortuitous accidents may suddenly change the entire disposition. I will, then, limit myself to the consideration and the statement of those rules which it is necessary to follow, and of the principles which must be respected to make preparations for the battle, and to distinguish the particular character which belongs to it. As to the manner of giving battle, there is nothing more variable. It differs with the nature of the operations to be executed, and the character of the mission the army has been charged with. It varies with the composition of armies, and the degree of genius belonging to the soldiers; still more, it differs by reason of the order of talent and the degree of capacity of the commanding generals.

I shall enter but little into technical details respecting the formation and preliminary position of troops, since these dispositions depend, above all, upon the nature of the ground upon which a general is called to engage the enemy. Thus, for instance, it is evident that a position near the battle-field, which may serve as a support and increase the means of sustaining ourselves, must be occupied in force, and in such a manner as to be fruitful of some salutary action, either when attacked, or when we ourselves assume the offensive. The strength of a position renders less disadvantageous the inferiority in the number of troops; defiles, placed in advance, render a part of our means of defence superfluous, and increase the difficulties of attack. As for the rest, the simplest reasoning, and oftentimes instinct alone, suffices to indicate the modifications necessary to be made in the manner of formation sanctioned by usage.

I will only recall, in a few words, that, aside from the influence of localities, as a fundamental principle the formation of troops upon several lines has been adopted. The first line is deployed, and the second is in column of battalions at a distance equal to that of the deployment, ready, if needful, to march or to move into line of battle; and a third line, composing the reserve, in column of brigades, ready to be moved wherever it may become most useful.

I shall make, however, one observation upon general dispositions. It is, that the command of troops should be divided in such a manner as to embrace both lines—that is to say, the corresponding parts of each should be under the authority of the same chief. The reason for this is easily understood. Since the second line is destined to sustain the first, the movements of the same fractions in both lines should accord perfectly. It is not the same with the reserve; it forms a complete and independent corps, whose means should be perfectly concentrated to act according to circumstances. Thus, a *corps d'armée* of four divisions, disposed to give battle, would have, in my opinion, the following formation:

In the first line, three brigades of three different divisions, and in the second the three other brigades of the same divisions, and the fourth division in rear, united, and formed in two masses, each constituting a brigade.

The cavalry would thus be placed: That of divisions, upon the flank or in rear of their respective divisions, and the great bodies of cavalry upon several lines and upon the flanks of the army, off the second line, and, in preference, upon that side where the country is most open and favorable to its movements and action.

As for the artillery, that of the reserve will keep in rear of the reserve infantry, ready to move wherever required.

Finally, I will add that the art of directing well a battle consists, particularly, in the judicious employment, and at the right time and in the right manner, of one's reserves; and that the general who, in a well-contested battle, has fresh and disposable troops at the end of the day, when his adversary has given his all, is almost certain of victory[1].

I will now establish the character of battles, by dividing them into two classes: Defensive battles and offensive battles.

For the first, the conditions of success are: The choice of a good position, whose flanks are well supported and the rear of which is perfectly secure and open to movements, with obstacles in front which render the approach of the enemy more difficult; lastly, brave and disciplined troops, commanded by an energetic and stubborn soldier.

Offensive battles, above all, require an excellent strategical combination and tactical skill, and, particularly, troops accustomed to manœuvre and to march well, and who are adroit and intelligent, and of a spirit not to be mistaken. It is necessary that the soldier should look upon success as a matter of course, and that he associate himself with it beforehand.

In applying these observations, which I believe to be of rigorous exactness, to the spirit of different armies, and by taking, for example, troops which resemble each other the least, we recognize the French as those troops which most nearly satisfy the requirements of offensive battles, and we accord to the English the first place in defensive battles. If it is, besides, remarked that the difficulties of administration and the maintenance of troops are immense in offensive warfare, while for the defence only money and the will are required; and when we further reflect that the English army, by its composition, manners, and wants, ought to be more abundantly supplied than any other—we are more and more strengthened in the conclusion that defensive warfare, with all its consequences, is more in accordance with the talents of the English army, and that it would be less easily conducted with a French army.

The events of the Peninsular war, yet fresh in our remembrance, demonstrate this truth. The English general, either by his nature and his peculiar character, or owing to his ability in comprehending the circumstances in which he had been placed, understood, from the beginning, the system which he ought to follow, and never deviated from it.

For a long time he perseveringly took advantage of a powerful auxiliary which the force of circumstances had given into his hands—our misery; he never ceased to turn it to account. His army, abundantly provided with everything, able to concentrate at any day, was constantly in a condition for movements, and always menacing. Military and political calculations alone were considered in its operations; while the French army, suffering from all kinds of wants, and discharging duties of every nature, daily lost in strength and means. If a position could not be attacked, the English general occupied it and waited until it was in danger of being turned, or until the French army had thrown itself against insurmountable natural obstacles, in pure loss consuming its valor.

Thus, when Marshal Masséna, at the head of a superior army, threatened to invade Portugal, Wellington took position in rear of two strong places, being, besides, covered by the Coa, and waited until the French army had wasted a part of its resources by two sieges—abandoning to the fate of war the garrison of these two places, which did not belong to his army—he retired when they capitulated, and fearful of being attacked, took a position at Busaco. After having repulsed the French army, which inconsiderately had attacked him, he retired and disappeared while the latter was manœuvring to turn him, and the English army withdrew within the lines of Lisbon, where art had added to powerful natural means of resistance[2].

The English general waited patiently until want and misery had disorganized and destroyed the French army; he followed his system in so rigorous a manner that he left it at peace, although it was within his sight and the range of his artillery, and unable to give battle or to oppose any serious resistance, weakened as it was by the absence of from fifteen to twenty thousand men who, leaving their arms stacked, dispersed over a space of some fifteen to twenty leagues into the interior of Portugal to seek for provisions. Nearly reduced to one-half, the French army returned to Spain, after having abandoned all of its cannon, its entire material, on account of the want of draught-horses, and with three-fourths of its cavalry dismounted. It had sustained immense losses, although it had but once fought—at Busaco—and had been engaged in only two combats of little importance during the retreat.

Wellington always followed a like system; and when later he was face to face with Napoleon at Waterloo, he again fought a defensive battle[3].

It is then seen that, in a defensive war, which is always a question of time, battles should not be given except very rarely—since marches and divers circumstances sometimes disorganize and destroy the means of an adversary more surely than the most signal victory.

Regarding the particular requirements of offensive battles, they should always be given by the front; and the talent consists in forcing the enemy, by means of wisely-conceived dispositions, to attack where we have been able to render resistance most easy. But there are likewise battles which, commencing with an offensive movement, are afterward reduced to a defensive action; this happens when prudent and circumspect commanders are at the head of nearly equal forces, ready to give battle.

The campaign in Spain in 1812 offers an example of this kind: the English army being superior by eight thousand infantry and four thousand cavalry to that of the French. The French general, after having for a long time remained on the defensive, awaiting promised reinforcements, and officially informed that they would not be sent, was obliged to assume the offensive to prevent the continually increasing dangers of his situation.

But in assuming the offensive, and in forcing the enemy to recoil by means of strategical movements, he did not wish, however willing to fight, to renew, by an inconsiderate attack, those events which had before taken place. He desired, on the contrary, should there be any battle, that it be fought upon ground of his own choice, and to accept, not to initiate it. On the other hand, the English general, true to his system, was equally desirous of reducing any action to the defence of his position. From it arose the remarkable movements which took place from the Duero to the Tormes toward the middle of the month of July, 1812.

This system being followed by both parties, the English army was obliged to make a retrograde march. Its return upon the Aguéda and reoccupation of Portugal would, incontestably, have been the immediate result of this part of the campaign, if a movement had not been executed without orders in the French army, and if the marshal who commanded it had not received a serious wound three-quarters of an hour before the battle, which unsettled the command and prevented the timely reparation of committed errors, and led to an action which should only have been brought about at a later day and under better auspices. Despite these drawbacks, the losses were equal in both armies.

Although I believe firmly that the French troops, when well commanded and properly provided for, are equal to all modes of warfare, I yet believe that offensive war is more within the spirit, nature, and the character of our soldiers. The same was, above all others, the particular genius of Napoleon.

I have already remarked that no man has ever possessed, in a higher degree than he, the talent of strategy; all his offensive marches, until the period of the Russian war, were skilfully conceived. The powerful resources of which he disposed, the energy with which they were employed, and the *morale* by which they were animated—his activity, the absolute liberty of his projects and combinations—all were calculated to precipitate events, and, by exalting the spirit of his soldiers, to overwhelm the enemy in advance with discouragement; and he was not very far from the fear of being conquered in one defeat. What series of splendid operations were executed by this almost supernatural genius! In the beginning of his career, in Italy, he turns all the positions of the enemy and beats him in detail, before the latter had been able to concentrate. He passes the Po without having an enemy in his front, because he had foreseen his movements. The war becomes defensive; but soon he changes its character, and, in attacking, he again applies the genius particular to him.

In 1800 he enters Italy and forces the Austrian army to receive battle in the most discouraging situation, and under most grievous disadvantages, after having lost both its communications and point of retreat.

In 1805 the simple direction of his armies which he threw upon the Danube in masses, after having occupied the Black Forest with the heads of his columns to engage the attention of the enemy, decides the question of the campaign—because Mack, instead of bringing about the catastrophe of the Austrian army by an unreasonable confidence, ought to have retired. As it was, one simple movement placed us in possession of the whole of Bavaria.

At Austerlitz we see a tactical movement decide the fate of the battle in a few hours. At Jena the same astonishing results were brought about by like means. As long as this system was followed, we see every enterprise of Napoleon crowned by equal success.

In 1809, at the outset of the campaign before Ratisbon, the same spirit regulates his operations. But soon his system changes. The passage of the Danube, after having been baffled the first time, is executed with success, and followed by the victorious battle in the plain of Wagram. Here an attack in front, a direct manœuvre, constituted the combat. Attending circumstances left no choice. The passage of a river like the Danube is not an easy matter, and can not be executed without the knowledge of the enemy; and if an army placed upon the opposite bank seeks to prevent it, we must at once prepare for a heavy engagement while debouching; then the rapid accumulation of means, and the energy with which they are directed, are the only assurances of victory.

In 1812 it depended upon his own will to give to the great battle which was fought upon the Moskowa the character of his preceding victories; by a simple flank movement he could have engaged the Russian army with much greater advantage, and a chance for still greater results in the future would have been opened.   But already a marked taste for direct attacks, and the enjoyment of employing force, appeared to manifest itself with him, and a sort of disdain for the *concours* of art and the combinations of the mind.   He vanquished, but with immense losses and inadequate advantages.

In 1813 he varies in his application of strategy.

At Lützen, being surprised, the battle commences by being defensive, but soon becomes offensive.

At Bautzen his strategical movements are well and skilfully conceived.

But at Leipzig we ask how Napoleon, who had it in his power to change the theatre of operations, could himself select a battle-ground so disadvantageous, and which the simplest calculations should have shown as being fraught with disaster.   The battle of the 18th of October was defensive, and presented no chance of success, since the battle of the 16th had not been gained; and as the enemy, on the 17th, had received a reinforcement of one hundred and fifty thousand men, he should have then avoided it, and should have retreated without delay.

The Battles of Brienne and of Craon, those of Laon and Arcis, fought in France, could effect no real advantage, owing either to the manner of concentrating the forces or to the direction which the attacks assumed. All operations of this epoch were limited to partial movements directed upon separate corps.   On these occasions only the remaining energy of the French army was properly exercised, and these combinations were, besides, within the range of Napoleon's genius—who, sundry times thereafter, made several happy applications, as at Champ-Aubert, Montmirail, Vauchamps, and Montereau—where, in giving to an obstinately-maintained defensive the character of the offensive, he made use of the greatest feature of his genius.

But at last, reduced to the necessity of giving battle, through the union of the enemy's entire forces, when obliged to deliver it he should have given it a defensive character, selected a position under the walls of Paris, and fortified that city; he should have concentrated all of his resources and those of the capital, which he alone was able to turn to account, and there, for a last time, tried his fortune.

If fourteen thousand men, consisting of remains of a former army, abandoned to themselves, could, in an open country—without a single work of art to sustain them, and deprived of all succor which the city would have been able to furnish, owing to the disappearance and the flight of the superior authorities—resist, for ten hours, the colossal forces united in their front, of which fifty-four thousand men had been engaged and thirteen thousand disabled, it is easy to foresee what would and should have happened if thirty thousand men could have fought under the protection of good works, tripling their force, and, aided by the resources of Paris, concerted action would have been the result of Napoleon's presence and authority.

But this sort of resolution was not within the range of his genius; he neither could bring himself to contemplate it nor to prepare for its execution; in this last resort he placed his reliance solely upon the powerful lever with which public opinion appeared to favor him, and, however mighty this power of opinion may be, it can only be durable as long as the condition of the country is based upon something positive and real.

One more remark upon offensive battles. At what hour should they be given? and this is a question of great importance, worthy of examination.

The hours, if the choice can be had, should be changed according to circumstances. If there is any decided superiority authorizing a firm reliance upon victory, the attack must take place early in the morning, so as to enable us to draw profit from obtained successes. There is not one military man who does not keenly remember the grief he experienced when the night closed upon success, or the impatience with which he expected the morning after a reverse.

Again. The attack should be made as soon as possible, when the troops are all held well in hand, and when the enemy has not as yet concentrated his. It is vainly asked why Napoleon, at Waterloo[4], during the longest days of the year, attacked the English only at eleven in the morning, though well knowing, through an intercepted letter of Blücher to Wellington, that the former would not be able to debouch until four o'clock in the evening; since, if Napoleon was victorious, he would have confronted the Prussian army after having beaten the English; and had the battle been against him, he at least would not have had a second army upon his hands in the very midst of the engagement.

Great military questions are nearly always reduced to simple ideas; and here the formula is, whether one has a better chance in fighting one against one, or one against two.

But if the forces are pretty nearly equal and render victory uncertain, it is better to attack toward the middle of the day; the consequences of a reverse are then less to be feared, and a general should, above all, think of preserving his army. The destruction of the enemy is only effected with the second line of the order which our duties and interests prescribe to us. And, furthermore, if the question rests unsolved, the whole night remains to prepare for a new attack and other combinations. Then the troops have had some repose, they have been able to take a meal before the engagement, and they are in a condition to display strength and energy. On the contrary, the army which defends itself

15

is full of reflection and agitation; it can not give itself up to the same degree of repose, and, oftentimes, its *morale* decreases as the moment of action approaches.

In the midst of our triumphs in Italy we experienced two slight reverses in two succeeding days—at Cerea and Alle due Castelli—owing to the extreme fatigue and a little disorder which existed in Masséna's division. As it was important to keep Wurmser shut up in Mantua, and to guard against a new check, the troops were permitted to repose until noon; they took arms only after having had a meal, and the victory of Saint-George's was not doubtful for one moment.

To sum up. Defensive battles are more a part of the profession, while offensive battles, well-prepared and well-conducted, are the portion of genius. Such was, likewise, the peculiar character of the wars of Frederic II, because the great defensive Seven Years' War had almost entirely the character of the offensive; and, in this respect, his campaigns resemble several of Napoleon's very much, with the difference resulting from the times and the state of the art.

In attentively perusing the recital of the actions of great generals, the character of their troops are recognized by the manner of their employment. Their particular excellencies are always detected, since it must be admitted that those who excelled in a particular kind of war had a special genius for it; the instinct which nature has imparted, if it be not our foremost guide, at least powerfully contributes to the development of our faculties.

In all centuries the operations of great generals have worn a peculiar physiognomy; even those conducted by men most frequently compared present essential differences to our reflection. The campaigns of Turenne and of the great Condé bear no resemblance whatever; it is the same with the generals of antiquity, if we compare Alexander with Cæsar and Fabius, or Hannibal with Scipio.

A skilful general should, at the beginning of a campaign, be fully impressed with the importance of his situation, the more or less favorable chances of which depend upon the character of his troops, their number, the task assigned to him, and the resources which have been placed at his disposal. The powers thus given him he ought to bring into action in the manner most likely to attain the object with which he has been charged, even if the particular mode of execution required of him does not agree with his tastes or wishes.

NOTES.

1. **Confederate Corps of Reserves.**—Before we may expect any extraordinary achievements from reserve bodies of troops, another principle must be adopted in their choice. Now we have no body of reserve corps in any of our armies. Certain regiments or brigades are simply detailed as a reserve corps for that

temporary duty. In doing so, we neglect one of the greatest stimulants to excel upon the field of battle: the creation and education of bodies upon whom the whole army looks with pride, and to whom superior excellence is willingly accorded; bodies, who, when they once march, are sure to carry everything before them, and whose organization, instead of being temporary, should be permanent.

Such bodies can only be created in times of war, because upon the field of battle their glories ought first to be incontestably established. Now the proper time appears to have come to create them. The bravest, best drilled, and the most steady troops should compose the corps of reserves of every army. We have all the elements in our armies; and we have an example in Napoleon's Guards, and in the present Imperial Guard of France.

Has a regiment particularly distinguished itself upon the field of battle, let the vote of the whole army assign it to the corps of reserves. The next battle-field would, no doubt, show great emulation, and every army would soon have a splendid corps of reserves. Has an officer particularly distinguished himself, let him be promoted into the reserves; and only the bravest general among the brave should command a brigade of reserves.

In the Italian war of 1859, one of the principal reasons which contributed to the speedy overthrow of the Austrian army was the want of an Imperial Guard such as the French army had. This guard is its reserve corps. In it all old veterans serve, and a young soldier has no chance to be admitted to its ranks before he has given proof of his prowess upon the field of battle.

With the Austrians the whole army consisted of comparatively young soldiers, but few having served more than two years, and none over three. It is, therefore, easily understood why they could not stand before the French Imperial Guard.

There seems to be no special legislation of Congress, or even any particular orders from the Department of War, necessary to effect the object (except such an order from the latter would at once effect it). Every general commanding an army can himself, and ought himself, to do it. Many of our indecisive fields of battle might have worn a different aspect but for a fresh corps of reserves. We are forcibly reminded of Sharpsburg, and, alas! too many other fields of that kind.

"Had we but had five thousand more men," may be heard after almost every battle. This has its significance, and expresses some general want.

2. The lines of Torres-Vedras, in Portugal.—The famous lines constructed here during the Peninsular war by Lord Wellington were not only, in their kind, the proudest monument of British military science, but present the most stupendous example of a mountain-chain of entrenchments which any age of the world has yet seen.

The recoil of Masséna's army from these lines formed the point of reaction in the career of French conquest, from which all the subsequent reverses of Napoleon may be dated. It was at the close of 1809 that these stupendous lines were commenced. The offensive movements which led to the Battle of Talavera having put to the test the value of Spanish co-operation, and having fully demonstrated the utter inefficiency of their armies from want of organization, want of discipline, and skilful officers, it became apparent to Wellington that the contest would, in the next campaign, devolve on the small body of veteran British and the newly-raised Portuguese troops under his command, and a defensive system of warfare ensue. To prepare for a final struggle was thenceforward the great object of consideration; and as the hope of successfully defending an extended and open frontier, like that of Portugal, against a very superior and highly skilful enemy could scarcely be entertained, it was decided to seek out some positions in the lower part of Estremadura, not liable to be turned or passed, and having an assured communication with the sea, which should command all the approaches to Lisbon, and which positions, being entrenched in the strongest manner, would offer a point of concentration for the whole of the defensive forces of Portugal, army, militia, irregular, etc., where they might, in conjunction with the British, be victualled and supplied with ammunition for any period of time, while occupying a most favorable field for deciding the fate of the capital and the kingdom in a general action. With these views, while the army was centred on the Guadiana, Lord Wellington, after a minute personal reconnoissance of the country, commenced a chain of fortified posts across the Peninsula. When completed they formed a double and nearly parallel chain of redoubts and other entrenchments. The outer, or advanced line, extended from the mouth of the small River Zezandra, on the ocean, through the mountain point of Torres-Vedras and Monté Agraça, the keys of the position, to Alhandra, on the Tagus; and following the trace of its defensive features, this outer line measured twenty-nine miles. In rear of this, the second or principal line of defence across the Peninsula, had its left on the sea, at the mouth of the little River St. Lorenzo (in front of Ericena), and its right on the Tagus at Via Longa—occupying, on its trace, the strong mountain-passes of Mafra, Montachigne, and Buccelas, through which run three of the four great roads to Lisbon, while the fourth skirts the river. The principal line, in its sinuosities, measured twenty-four miles; the direct breadth of the nook of the Peninsula, between the flanks of the two lines, being, however, twenty-five and twenty-two miles respectively.

3. **Wellington's Strategy.**—As a matter of interest, we subjoin here the remarks of Lieutenant-Colonel Graham, of the English army, on the strategical talents of the Duke of Wellington, mostly condensed from General Jomini:

"In respect to the influence produced by Wellington on the progress of the art of war, although his campaigns tend to confirm the great truths of the science, it can not be said that he created or brought into practice any important principles in the higher branches of the science which had not been previously illustrated in the operations of Napoleon, the Archduke Charles, and other celebrated generals.

"At the same time it may be claimed for Wellington that he had a system of his own, which forms an important and interesting subject for reflection and discussion.

"The combination of the defensive with the offensive, generally used by Wellington, has been thus described by various military writers:

"In the Peninsula, and more particularly in Portugal, he had under his command a mass of native troops, better adapted to act as light troops and harass the enemy in his operations, than for a pitched or regular battle.

"Having experienced the fiery ardor of the French attack, the impetuosity of columns led by Masséna and Ney, he devised very skilfully the means of first blunting the force of that impetuosity, and then overcoming it.

"His plan was to choose positions very difficult of access; the approaches he covered with clouds of Spanish and Portuguese tirailleurs, thoroughly conversant with the art of taking advantage of inequalities of ground; his artillery he placed partly on the tactical crest of the position, partly a little in rear of it.

"By these means he shattered the columns on the march by a murderous fire of artillery and musketry, while his excellent British infantry, one hundred paces behind the crest of the hill, was kept in reserve without being exposed.

"When the columns reached the summit, fatigued, out of breath, and already decimated, a general discharge from the infantry and artillery of the second line burst upon them, followed up immediately by a charge upon their half-disordered masses.

"This description has only reference to the campaigns of Wellington in Spain and Portugal—rugged countries to which, as well as to the peculiar characteristics of the troops composing his army, such a system was perfectly suited. In the campaign in Belgium a modification of this system was necessarily adopted.

"The position at Waterloo was on a plateau, with a gentle slope in front of it, forming a glacis, over which the infantry had a magnificent range of fire, and its effects were terrific. The flanks of the British were well protected, and Wellington could observe the movements of the enemy from the summit of the plateau, while his own were, in a great measure, concealed. With all these advantages it still remains a question whether his system would have been successful on that day, if a multitude of other circumstances had not lent their aid.

"We admire the sagacity with which the various materials, or nationalities, placed in Wellington's hands, were used to the best advantage; but one of the most striking characteristics of his operations, as far as respects the art of war, is his almost constant use of deployed lines two deep.

"His successes in the Peninsula and at Waterloo, gained with the troops so formed, and generally attributed to the deadly fire of his lines of infantry, have given rise to controversial discussions on the relative advantages of the formation of troops in columns and in lines.

"In these discussions, highly useful in some respects, the tendency in too many instances has, unfortunately, been to reduce the system of war to absolute rules; as if it were possible to decide absolutely that troops should on all occasions be formed either in column or in line.

"The locality in which they are to act, national character, and other circumstances, must be considered by the general; and it is for his genius to decide when and where the one formation or the other may be the most advantageous."

It is curious to perceive how the paragraph commencing with "The position at Waterloo," etc., differs from the same in Note No. 4.

They are both translations from Jomini. Of course Mr. Pardigon's is the correct version, and Colonel Graham either did not know French sufficiently to translate correctly, or he purposely perverted the meaning of Jomini.

4. The Battle of Waterloo.—Of the Battle of *Mont-Saint-Jean*, as it is called by the French, *La Belle Alliance* by the Prussians, and *Waterloo*, very singularly, by the English—the Village of Waterloo having been fully two miles to the rear of the English front of battle—so much has been written, and so many diverging opinions by many celebrated military writers are on record, that it appears exceedingly difficult to form a just and correct estimate of it as regards the merits of the great leaders therein engaged. And while we know, in almost every battle fought by Napoleon, to whom to ascribe the honors of the victory, to this very day English and Prussian writers violently dispute who really gained the Battle of Waterloo; and the latter proclaim Marshal Blücher as the victor as firmly as the English maintain that to Wellington alone the honors are due. They both, however, agree that the French were beaten by the skill of their own generals.

But to throw some light, even at this late day, upon the subject, I beg to refer to the incontestable fact of numbers, as established in the official data of "Captain Siborne's Waterloo Campaign." We find there evidence, which perhaps has

15*

as yet but little been considered, that out of 67,661 men under Wellington present upon the field of battle, but 23,991 were Englishmen (and it may be well supposed that the greater part of those were gallant Irishmen), and the remaining troops—that is, 43,670 men—were mostly Germans, in which number are included nearly 18,000 of Dutch and Belgians. Of artillery, these foreign troops actually had *one-half*—that is, seventy-eight out of one hundred and fifty-six guns.

Furthermore we see. from the same source, that the loss of the English troops, not including that of the large body of foreign troops, was 5,572 in killed and wounded during the entire action—that is, from morning, eleven o'clock, when the village clock of Nivelles gave the signal to the bloody conflict, until night covered the ghastly scene—and that the Prussians, who appeared upon the field of battle late in the afternoon, still lost as many as 4,810 in killed and wounded; so that we must infer, if the English troops fought very splendidly, the Prussians did not the less so. We see, therefore, no substantial grounds to award the greater glory to either.

But, despite the convincing figures of Siborne, it must be a matter of astonish. ment if the English, far from according any credit to these large bodies of foreign troops, both military men and historians have not only depreciated their services, but also not hesitated to cast injurious reflections upon their character—a course which can not be defended, since there is no record that those troops failed in their duty; on the contrary, we know that a body of these very auxiliaries defended and maintained the important position of the Orchard of Hougoumont against the severest efforts of the French. It illustrates strongly a feature of their national character—great selfishness and unblushing unfairness whenever their national pride, interest, or prejudices dictate a course different from that which justice, honor, and magnanimity would appear to indicate.

In regard to the respective merit to be attributed to the great commanders engaged in the Battle of Mont-Saint-Jean, so much is established, that Napoleon's combinations, which culminated in this battle, admit of no adverse criticism ; they are acknowledged, even by English historians, as having been signalized by the genius so singularly revived after the great reverse of Leipzig, in October, 1813, and which was demonstrated by his subsequent magnificent campaign of 1814, when, in the plains of La Champagne, he, with barely 60,000 men, kept in check, for a long period, the combined hosts of the Russians, Prussians, and Austrians, amounting to over 200,000 men.

Not very long ago two officers, who speak the French language—one, Colonel Charras, a French exile—the other, a Captain Brialmont, of the Belgian army—have written voluminous works to prove that Napoleon showed very little if any generalship, extolling that of Wellington to the skies. But their motives are easily perceived, and their reward has been the contempt of all unprejudiced men and the high praises of the British press.

The emperor's plan of operations, upon setting out from Paris, was to anticipate the junction of Wellington with Blücher; to turn to the right and beat the Prussians, and then seek Wellington and defeat him. The first part of his intended operations succeeded as well as could have been expected, and Blücher was beaten at *Ligny* on the sixteenth of June—but two days anterior to the final struggle—and it would have ended in the total destruction of the Prussians but for the unpardonable want of vigor and resolution on the part of Ney, who, having been directed to march by way of *Quatre-Bras* and *Bly*, upon Blücher's rear, permitted himself, despite the emperor's urgent despatch—"*The fate of France is in your hands*"—to be kept at bay by a few thousand men in front of Quatre-Bras—giving Wellington time there to concentrate and defeat him.

Both Wellington and Blücher now retreated; the former from Quatre-Bras to Genappe and Waterloo—the latter from Ligny to Wavre. Wellington halted in front of the straggling little Village of Waterloo to make a stand, Blücher having promised him to unite on the 18th.

To prevent the junction of their forces Napoleon had detached Marshal Grouchy, with a force of thirty-five thousand men, toward Wavre and Sombref, with strict directions to watch and keep within sight of Blücher. Either a traitor or an imbecile, but most likely the former—since his previous glorious career gives him greater claims to intelligence—the marshal failed to do this; and, although within sound of the firing of Waterloo, knowing that the fate of his country was about to be decided, and against the urgent entreaties of Count Gerard, one of his lieutenants, he refused to march to the emperor's assistance. In this circumstance we must seek the main reason of Napoleon's loss of the battle—for the issue could not have been doubtful had he been able to dispose of Grouchy's corps. When, finally, after having worsted the English upon every point, a corps of seventy-two thousand Prussians, with two hundred and twenty-eight cannon, made its appearance upon the field of battle, it was beyond human power to preserve the victory. Still, the well-contested action is shown by the great loss of the Prussians, whose main body was not engaged before seven o'clock in the evening.

From the battle-field of Waterloo, or rather from the Chateau of Hougoumont, it is about ten miles north to Brussels, six miles south to Genappe, eight and a half miles south to Quatre-Bras, fifteen and a half miles south-east by east to Ligny—all on the same road—and but eight and a half miles east to Wavre—which distance could have been made by Grouchy in a forced march of three hours, by his infantry and foot-artillery, and in one and a half hours by his cavalry and light-artillery.

On the other hand, it is an equally well established fact that Wellington was already beaten, and, but for Grouchy's failure, would not now be called the great victor of Waterloo and conqueror of Napoleon. General Jomini gives the following as the causes of the loss of the battle in a chapter written in 1856 (*vide* p. 206 of Mr. C. F. Pardigon's translation of "The Practice of War," Richmond, 1863):

"At Waterloo, the English general took position on a plateau sloping gently down and forming a glacis, where the artillery formed a magnificent line of fire, and ranged with its most terrible effect; moreover, both flanks were well protected. Wellington could descry from the crest of the plateau the least movements of the French army, while the latter could not see anything of his. But, notwithstanding all those advantages, his system could not have prevented him from losing the battle, had not several other circumstances turned in his favor.

"1st. The rain soaking the ground, rendered very slow and toilsome the offensive march of the French, deprived of all the impulse of the first attacks, and did not permit them to be properly supported by artillery.

"2d. The formation, from the beginning, in too deep columns, especially at the right wing.

"3d. The incoherent use of the three arms, since the cavalry and the infantry made several alternate attacks, without ever being simultaneously engaged.

"4th. Lastly, and *above all, the unexpected arrival of the whole of the Prussian army, falling at the decisive moment on the right flank and* NEARLY ON THE REAR *of the French.*

"All military men of experience will agree that, in spite of mud and the good bearing of the English infantry, if the main body of the French infantry had pushed forward in *columns by battalions, just after the great charge of cavalry*, the combined army would have been broken and thrown back upon Antwerp. Even independently of those circumstances, *the English army, but for the arrival of Blücher, would have been compelled to fall back!*"

Alison and other English historians admit that Wellington was outgeneralled in the beginning of the struggle. Wellington, whose claims as a master of defensive warfare can not be denied, can, however, not be ranked as a first-class general; least of all could he be placed in a line with the incomparable genius which controlled the operations of a Napoleon, as the English would make us believe. General Jomini tells us, in Nos. 2 and 3, of Napoleon's failures—and they were serious; still Wellington did not take advantage of them, merely maintaining his defensive position. It may be safely asserted that, in like circumstances, had Napoleon seen that Wellington committed such mistakes, he would have acted widely different. Furthermore, in this battle, which alone would give to Wellington not even the reputation of a second-rate general, he violated one of the most important maxims of war, by taking a position in front of the woods of Soignes, with but one military road in his rear to retreat upon, thus leaving behind him a defile. This road was at that time the only one practicable for artillery, cavalry, and baggage-trains. There are certainly, as all English writers maintain, in order to shield Wellington from any censure, many roads which traverse the forest of Soignes; but they were, at the time, mere foot and bridle paths, upon which a retreat could not be conducted without destruction. Had the Prussians not come up in the nick of time, Wellington would have been totally routed upon his retreat; and the consequences of his violation of one of the principles of war would have become more disastrous to him than the passage of the defile of Hohenlinden was to the Austrians fifteen years before. If, however, his partisans still contend that Wellington acted right in thus placing himself—a defence which could not possibly be supported by any rules of the art—the inference is not improper that he was driven to bay by Napoleon's capable operations, which resulted in the Battle of Ligny.

Wellington is the only remarkable Englishman the wars with Napoleon produced. The immense consequences which followed in the train of the Battle of Waterloo could not but heighten the lustre of that victory—and to these two reasons the great admiration the English possess for their Wellington can easily be traced.

As to Marshal Blücher, history has pronounced its judgment upon him likewise. He was a general personally very brave, and always ready to engage in combat himself, rather than to employ his head—leaving that disagreeable work to his generally able chiefs of staff. The obstinacy with which he maintained a combat, and his great loyalty, so little found among French generals, were not the least of his good traits, and under a proper leader he would most likely have made a general of the character of Murat, but not a Ney. His often injudicious dispositions and inconsiderate impulses led to more defeats than victories. His coarseness, however, was astonishing, and almost amounted to brutality, though he could have never become a Butler.

Under his explicit orders the Prussian squadrons pursued and slaughtered the French with a degree of barbarity that has but seldom been equalled in history, and which, long since, has been stamped with the seal of unqualified condemnation; and the remembrance of their cruel deeds will be fresh in the breasts of Frenchmen as long as one of their national standards is raised.

Blücher's motto was: "*Forward;*" but not the "forward" which aims to achieve some great and humane end through the medium of military operations, but that which is characterized by rapine, murder, and annihilation—which, at last, would bring us back to the horrors of the most uncivilized times, and which to check is the duty of every Christian nation.

To apply somewhat the cause of the loss of the Battle of Waterloo to our own struggle, instances could be cited where generals, charged by the general-in-chief of the army with which they served with certain operations upon which the fate of the battle, or the campaign, or, who knows, the cause, might have depended, appear to have signally failed to carry these operations into effect. They permitted the struggle to proceed, and did nothing, when their appearance upon either the flank or the rear of the enemy might have secured the most decisive victory; or, when the commander of a large detachment, though within hearing of the conflict, and knowing that the great struggle was going on, remained indecisive, did not move to where the fire was heaviest, and had either not the courage or the good sense to assume the responsibility of hastening to the support of the main body.

There are omissions in military life which may not be within reach of the Articles of War, but which, sooner or later, will bring upon the officer thus failing the condemnation of the entire nation, and which will consign his name to those who ignominiously deceived the high trust reposed in them. Such failures can be excused as little in a simple colonel as in the exalted rank of a general of whatever grade, because the FATE OF THE COUNTRY, and, in our case, *the immense stake of human liberty*, may depend upon it. Grouchy, whether he was in reality a traitor, or whether he failed simply to seize the decisive moment, is held in execration. No Frenchman mentions his name but with a curse upon his lip or in his heart. Such will be the fate of those of whom hereafter it shall be established that they betrayed their trust; that they permitted their brave comrades to be slaughtered, and a cause to be ruined, despite the wisest and most effective measures of both the general-in-chief and the government, when they had it in their power to turn the scales of victory, and bring about the glorious consummation of our struggle.

CHAPTER VII.

CONDUCT OF THE GENERAL THE DAY FOLLOWING THE VICTORY.

Many know how to gain battles, but few how to profit from them—Losses in a battle increase indecision and timidity—Schœrer at Loano—Clairfait at Mayence—Moreau in 1800—Brune on the Mincio—Napoleon's conduct—He never needed to fight a second battle—Great fatigues momentary—Abundance and repose gained for a long time—Limits—Russian campaign—An offensive movement of three months—Requirements of a serious pursuit—Necessity of suiting the means to the end—After Wagram—March upon Znaïm—Passage of the Taya—Offered succor must never be refused.

NOTE.—Marshal Davoust.

Generals who gain battles are less rare than those who know how to profit from victory. They would make the engagement the object, while it is but the means. This may especially be remarked of former wars; still, even in our own times, examples are not wanting.

An ordinary general thinks only of the losses he has sustained, and hardly suspects those of the enemy; from it arise indecision and fatal timidity, instead of a feeling of confidence, most probably authorized by all attending circumstances.

In 1795, after the Battle of Loano, Schœrer could have invaded Italy without any serious engagement. In the same year Clairfait, after his signal victory before Mayence, could have easily advanced under the very walls of Strasbourg, had he marched without delay. In 1800 Moreau would have been able, by means of rapid movements, to complete his successes at the very outset of the campaign. In the same year, in Italy, Brune, after the passage of the Mincio and the Adige, could have entirely destroyed the Austrian army which was retiring before him; the least display of energy would have been sufficient, to such a degree did circumstances favor him.

Napoleon is the first general, in our epoch, who knew how to draw from victories all the results of which they were susceptible. After a battle had been gained he marched with rapidity to pursue the enemy, in order to gain easy successes, and to deprive him of his remaining little confidence. With such a system, a new battle became rarely necessary for the attainment of an important object.

While thus marching it was, of course, impossible to provide in any great degree for the wants of the army, and inconveniences resulted therefrom, but they were of much less weight than the advantages which were assured. As a fertile country was, besides, traversed, in the midst of a compact population, the sufferings of the soldiers were much moderated, the rapidity of the march was soon slackened, and important spoils, with immense resources, fell into the hands of the conqueror. Abundance and repose thereafter furnished the means of not only repairing the losses, but also of augmenting the effective. While Napoleon made war in Germany he acted in this manner, and found himself well off. Vienna, twice occupied, supplied him with incalculable resources, and was itself a pledge in the negotiations, whose value could not be too highly estimated.

But there is a limit in these operations which can not be exceeded without impunity. When this system of war was applied to Russia, the question was not to make ten or twelve rapid marches, in the midst of a country full of resources and inhabited by a mild people, accustomed to order and obedience; it was, on the contrary, an offensive movement of nearly three months' duration*, almost without halting, in a poor country, offering but the most indifferent resources, and inhabited by an oftentimes hostile people. This movement had not for its object the

---

* Passage of the Niemen, 23d of June; entrance into Moscow, 14th of September —the movement occupying eighty-three days.—*Note of Author*.

pursuit of a vanquished army, but to overtake an army which was continually falling back upon new resources, while we were expending ours upon the march alone, and to alleviate the sufferings, of every description, which our soldiers underwent*—sufferings which soon engendered a state of things bordering on disorganization. Napoleon was then marching into certain misfortune.

If we are right in saying that, to satisfy one of the great principles of war, a general should seek to profit from his successes, and to neglect nothing to complete them by rapid movements in the pursuit of the beaten enemy, we likewise perceive that this rule is by no means without its limits, and that its application should always be subordinate to the exigencies of particular circumstances.

But if a serious pursuit must be undertaken, the most powerful and compact resources must be lavished upon it, suitable to surmount all obstacles. If, on the contrary, a general is forced to keep back, he strengthens the *morale* of the enemy, and advantages escape him upon which we could have rightfully counted.

After the Battle of Wagram, on the 8th of July, Napoleon ordered me to the command of one of the vanguards of the Grand Army. Masséna was following the main body of the enemy's army, which was retreating upon the road of Hollabrun. I was thrown upon the road to Nicolsburg, to pursue Prince Rosemberg, who was marching in that direction, and Marshal Davoust received orders to sustain me with his corps. I overthrew the troops in my front; and as they changed direction and abandoned the road to Moravia to march upon Laa, to effect there a passage over the Taya, so as to unite with their main body at Znaïm, I took the same direction. On the evening of the 9th, when I had arrived upon that river, I received a message from Marshal Davoust, written from Wilfersdorf, to the effect that if I wished succor he was ready to join me. I had experienced but little resistance from the enemy, and no circumstance authorized me in deeming any succor necessary; I, therefore, extended no invitation to that effect to Marshal Davoust.

I thought, besides, that since nobody had moved upon Nicolsburg, Davoust ought to cease marching upon that town. Still everything happened differently from what circumstances had authorized me to believe. Davoust marched upon Nicolsburg in order to subsist more easily; and I marched upon Znaïm, where I expected to meet nothing but a rear-guard, and to be able to join Masséna. But the retreat of the enemy had been slower than I thought, and two-thirds of his army

---

*The first corps, when entering upon the campaign, was eighty-five thousand men strong; at the review of Moscow it had only fifteen thousand.

The French cavalry of the line, when commencing the campaign, counted fifty thousand men; at the review of Moscow, six thousand.—*Note of Author.*

were still on the other side of the river with nearly the whole train, and one-third was in my front. I took a defensive position to resist his assaults; and this position being sufficiently near to Znaïm, influenced the retreat of the enemy's army, by commanding the passage of the bridge over the Taya. Despite his reiterated efforts, he could not dislodge me; still I did not feel the fault I had committed the less, in not calling for Davoust's support, and that which he himself had committed in not at once marching to sustain me. The retreat of the hostile army having been broken, and the greater portion of his forces obliged to retreat upon very difficult cross-roads and to again ascend the Taya, the sequel would probably have been the loss of the greater portion of his material and his disorganization. This success would have had incalculable consequences.

Succor which is believed superfluous should never be called for; but if any is offered it ought never to be refused. Often a fortuitous occasion arrives when, unexpectedly, such succor is of the utmost importance.

NOTE.

Marshal Davoust, Prince d'Eckmuhl.—This distinguished soldier of the French Republic and the Empire was educated for the army at the Ecole Militaire. At fifteen years of age he received his commission as a sub-lieutenant, and in 1788 was appointed to the cavalry in the Royal Champagne regiment. Soon after the breaking out of the revolutionary war he was raised to the rank of provisional *chef de brigade*, and in 1796 was general of brigade. In these capacities he fought in Belgium, on the Rhone, and Moselle, and in Egypt. He particularly distinguished himself at the passage of the Rhine, in the attack and capture of the entrenchments of the enemy at Diersheim and Honnau; in the expedition to Upper Egypt with General Desaix, and at the Battle of Aboukir, on July 25, 1799. Napoleon Bonaparte, discovering his military talents, appointed him a general of division, and gave him the command of cavalry with the Army of Italy in 1800, and in the following year appointed him inspector-general of the French cavalry. When the camp was formed at Bruges in 1803, the command-in-chief was conferred on Davoust. He subsequently received the bâton of a marshal of the empire, and in that exalted capacity commanded the *corps d'armée* on the Rhine and in Germany, contributing, by his skill and bravery, to the subjugation of Prussia and Austria. In 1808 he was created Duc d'Auerstädt, and in 1809 was raised to the dignity of Prince d'Eckmuhl. After the campaign in Germany he held command in the Grand Army which struggled for French dominion in Spain and Portugal. Upon the restoration of the Bourbons he was deprived of all the appointments he enjoyed; but in August, 1817, the privileges and emoluments of a French marshal were restored to him. He lived to enjoy them until 1823, when he died in Paris, on the 1st of June.

CHAPTER VIII.

RETREATS.

In presence of a stronger force—Delicate operation—Difficulty consists in the *morale* of troops—Different effect upon the soldier when he marches from the enemy—To scorn the enemy is to excite his respect—First dispositions—Necessity of retarding the enemy's march—Echelons—Partial combats—Rear-guard—Importance of artillery—The enemy kept at bay—1814, Blücher before Vindé—Six thousand men retire in peace before forty-five thousand—Retreat upon the Voire after the Battle of Brienne—The enemy attracted at Perthe—The defile of Rosnai passed without disorder, and as if on drill—Retreats with troops equal in number to the enemy's—Position chosen in advance—Calculation as to time, distance, and celerity—Retreat upon the Duero—Moreau retreating upon the Rhine—Masséna in Portugal.

Note.—Marshal Masséna.

That retreats executed in the presence of a superior enemy have always called forth great eulogiums is very reasonable, since they are among the most delicate and momentous operations of war.

The principal difficulty is to be found in the *morale* of troops, which, under such circumstances, is greatly shaken. It is a singular fact what a different impression is made upon the soldier when he stands face to face with the enemy, and when his back is turned toward him.

In the first case, he only sees what really exists; in the second, his imagination never fails to magnify the danger. It is, therefore, the duty of a general to inspire his troops with sentiments of pride and with confidence, justified in view of their situation, and to make the power and necessity of these impressions felt as one of the means of their safety.

He must teach the soldier that, by scorning the enemy, the latter will respect him. In ordinary cases, when a general is obliged to retire at the approach of the hostile army, and when no additional circumstances may require his prolonged stay in the position he is about to evacuate, both reason and prudence require of him to begin his movements before the arrival of the enemy. In keeping an interval of from six to seven miles at least, his march will be pursued with increased ease and facility. But circumstances may occur when it becomes necessary, above every other consideration, to retard the enemy's march, and to oblige him to lose time in making dispositions for attacking us, which will suddenly become superfluous, because we withdraw as soon as the prospects of the opening of the engagement becomes serious. In such an event both excellent troops, and great precautions on the part of him who commands them, are required. Security then must be found in the dispositions of the echelons and the precision of movements.

16

If the retiring corps is of such disproportion with the pursuing as to forbid the commander of the former to offer battle, partial combats, if prudently sustained and initiated, are still in his power and without danger. To this end, all necessary movements and dispositions should be made beforehand, so as to prevent his troops from being seized by any kind of embarrassment, and to maintain their march in as easy and light a manner as possible. He will strengthen his rear-guard with a sufficient body of artillery, without encumbering the same. The artillery should be well served, perfect in its material, and ought to contain several pieces of heavy calibre. This artillery, if divided into two or three sections, in echelon, will march with facility, and be well prepared for any successive and instantaneous stands for resistance that may be requisite. The enemy will thus be forced to stop and make his dispositions before the attack takes place, and at the moment his dispositions are finished the movement is once more begun, and we disappear. Then the enemy advances again, but is held at bay by the fire of the artillery, which soon will show its superiority over his—because the one, in pursuing, extends his columns, while the other party, by retiring, is constantly withdrawing from the battle-field and approaching the reserves.

And, from this time, a like constant change will take place in the respective forces thus coming into contact.

On the 25th of February, 1814, I executed a movement of this kind very successfully. I operated upon the left bank of the Aube, and my corps was composed of about six thousand men of all arms. The Prussian army, commanded by Marshal Blücher, and forty-five thousand strong, passed that river at Plancy, and marched against me. I took position upon the heights of Vindé, in rear of Sézanne. The appearances were such that I imparted to the hostile general the belief of being resolved to engage him. He made complete dispositions to attack and surround me, and massed a battery of thirty guns. The moment came; all my troops gave way with order, unity, and celerity, and the enemy began to pursue; but during the march, which continued all day, circumstances conspired in keeping him always at a distance, and he was obliged to stop his march frequently, in order to reassemble his forces whenever they became too pressing. I arrived at Ferté-Gaucher, while continually exchanging cannon-balls, and took position in rear of Morin. I had only lost those struck by the enemy's balls, and neither a single living man nor one piece of cannon had been left behind.

On the day succeeding the Battle of Brienne I was charged by Napoleon to retire upon Voire, and to take, at first, a position at Perthe, to attract the attention of the enemy the longest time possible, thus making a diversion in favor of the main body of the troops, who were retiring upon the Aube by the bridge of Lesmont. After parading my forces during the day, and preparing my retreat to make it with surety, I executed it, without any loss, under the cannon of the enemy. I passed the defile of Rosnai, without disorder and as if upon drill, before me the army of the enemy, which almost entirely had been directed upon me; it was unable to clear the Voire, the passage of which it had several times attempted in vain.

If the retiring army is of sufficient force to measure itself with the enemy, the dispositions will be analogous. Its security rests then in the manner of disposing the echelons; the end, however, is always the application of the fundamental principle already established—that is, to be always more numerous than the enemy at the moment of the engagement upon the battle-field.

The best disposition to be made in a similar combination, is as follows: To set out with the army at an early hour, leaving a strong rear-guard, which marches as late as possible, without, however, compromising itself, and to take a position upon a defensive ground at such a distance that the enemy could not arrive there before three hours of sunset. Whatever his ardor to engage, there is no time to make the requisite preparatory dispositions; and if he ventures upon an attack without having completed them he ought to be crushed, because the encamped army has all of its forces assembled, while he necessarily can only dispose of a part of his own.

It was thus when the Army of Portugal retired, in 1812, very inferior to the English, within sight of the latter, from the banks of the Tormes, to take a position upon the Duero, the enemy making no effort at any time to pursue it.

In 1796, when General Moreau evacuated Bavaria to retire upon the Rhine, and was followed by the Austrian army, he reduced the above theory to practice; and when, finally, too heavily pressed, and marching perfectly concentrated, he made a halt, gave battle, and came off victorious.

But if an army upon retreat, and even a simple rear-guard, finds upon its route an impregnable position, which the enemy can only seize by turning at a distance, it should always remain in occupation of the same as long as it can be done without danger. If the enemy manœuvres to cause its evacuation, the operations of the campaign will thereby be delayed, and time is everything to him who defends himself. If the enemy, in his impatience, attacks with some degree of ardor, and rushes upon obstacles of some extent, an easy victory will be achieved, sometimes exceedingly murderous for the enemy, and susceptible of changing the respective moral state of both armies in favor of the defenders.

This happened in Portugal on the 27th of September, 1810. The English army, inferior to the French, occupied a post on the 26th, upon the mountain of Busaco, a counterfort of the Sierra d'Accoba. The right of the position being impregnable, barred the road, while the left, supported by higher mountains, was of easy access. Masséna, to whom the emperor had recommended to profit by his superiority, and to force the enemy to receive battle, resolved to attack him at once, and, unhappily, without having before sufficiently reconnoitred the entire front of his position. After unheard-of efforts the corps of General Reignier succeeded in scaling the mountain under the enemy's fire; but having arrived upon the plateau, and finding the entire English army in line of battle, it was easily overthrown, and in a few minutes lost the ground which it had cost a painful and brave struggle of one hour to gain. Six thousand men were disabled. The next day, at the sight of the French army operating against its right, the English army disappeared. The result of this unfortunate combat changed the *morale* of the two armies; it diminished, on our part, that blind confidence which is so necessary to success, and enhanced that of the enemy. Had this event not taken place, an attack upon the lines of Lisbon would, most probably, have been attempted, and had it succeeded, the Peninsular war would have been terminated.

NOTE.

Marshal Massena, Prince of Essling.—Originally a private soldier, he advanced gradually, between August, 1775, to September, 1784, through the various ranks of corporal, sergeant, fourrier, and adjutant sous-officer. In 1789 he received leave of absence; but we find him rejoining the army in 1791, and obtaining the command of a battalion in 1792. In the autumn of the following year he was made a general of brigade, and in the winter a general of division. He made all the campaigns of the Revolution with the Armies of the Alps, Italy, the Rhine, and Helvetia. In 1799 he was commander-in-chief of the Army of the Danube. Everywhere so much success attended his operations, that he acquired the appellation of "*L'Enfant gâté de la Victoire*." On the assumption by Napoleon of the imperial crown, Masséna was made a marshal. The years 1805 and 1806 saw him commanding armies in Italy, and in 1809 he was at the head of the corps of observation on the Rhine. The failure of Soult, Victor, Ney, etc., induced Napoleon to send Masséna into Portugal, but it was only to sustain defeat at Busaco and Fuentes d'Onore, and to be compelled to retire from before the formidable lines of Torres-Vedras. Upon his retreat he lost some forty thousand men. Returning to France, he was made Governor of Toulon in 1813, chief-commandant of the eighth military division, and governor. In July, 1815, he was Governor of Paris, and died April 4, 1817.

CHAPTER IX.

NIGHT ATTACKS AND SURPRISES.

A surprise a good fortune—Good troops never meet with surprises, and only under-
take them—Night attacks—Opportunities—Few troops to be employed—Disorder
in the enemy's ranks—Certain success against indifferently disciplined troops—
Necessity of very precise instructions—Examples: Austrian army at Hochkirch
—Affair at Etoges in 1814; surprise of Ouroussow's corps—Chevert at Prague—
Prince Eugene of Savoy at Cremona—Details—Presence of mind of French soldier—Attempt of the English against Berg-op-Zoom; fine defence of General
Bizanet—General Laudon at Glatz—Enterprise against the T of Mantua—Colo-
nel Dufour at the Fort of Bard—M. de Sénarmont.

Notes.—1. The Battle of Hochkirch. 2. Celebrated turning manœuvre of Frederic after the battle.

A theory treating of surprises can not be established. It should be
an impossibility to execute surprises in the daytime, and it would always be so if every chief and every soldier did unceasingly perform
the duties of his profession with exactness and intelligence; but sometimes things happen differently. Still, whenever we succeed, and the
enemy can be surprised, we must look upon it as a fortunate occurrence,
from which too much profit can not be derived, since there is no operation in war promising a more prompt and easy consummation.

Troops which are in the required order of formation for such an enterprise, and who know that they are about to be engaged—who are, furthermore, imbued with a consciousness of their strength and a feeling
of confidence—such troops, attacking a surprised enemy, can not be resisted by anything; they have all the advantages on their side, and
have a right to consider the victory as already secured.

Only very good troops, animated with an excellent spirit, and commanded by a skilful general who takes his resolutions promptly, may
sometimes escape a catastrophe, when placed in like circumstances.
But it is also true that similar troops and a general of that kind will
never conduct themselves so as to be placed in such a condition.

It is entirely different with night attacks; there can be no surprise,
in the common acceptation of the term; in that case it is a mere unforeseen, impetuous attack, and ignorance of the true dispositions of the
enemy, owing to the fact of our inability to reconnoitre the presence
of the enemy during the night-time, except at a very short distance,
whenever armies are very near each other.

Only in case of very great proximity I believe such an enterprise to
be possible; because, otherwise it would be necessary to march a long
distance before the attack can take place, and then the different columns, at the moment of action, would most likely be unable to act with
any degree of harmony between them.

16*

I repeat, then, that such actions can only happen when two armies have approached very close to each other; then, bodies of medium strength alone should at first be employed; they should attack simultaneously upon several points, and seek, above all, to create disorder in the ranks of the enemy. If they succeed in this, the effects of a victory are obtained without any great sacrifices, and we are enabled to profit from it in proportion as the condition of affairs would thereafter permit us to do.

Against mediocre and indifferently disciplined troops such actions will especially be advantageous. If such troops, in the midst of the uncertainty of a real attack, are moved forward, confusion soon spreads among them, and sometimes it will even happen that the different columns mistake and combat each other, entirely to the profit of the assailants, who will be mere spectators of the scene; while he who attacks, on the contrary, employing but part of his troops, after having given to them precise instructions as to the sphere in which they will be required to operate, and having made them acquainted with the position and direction of the other columns, will be much less apt to fall into like fatal errors. More than once the columns of the same army, operating during the night, have each mistaken the other for the enemy, and much damage has been done*. If a simple accident can lead to such actions, it is easily understood how possible it is to originate them; they will then be much more serious, since the presence of the enemy is a reality, and he can make himself felt in a direct manner. It is, therefore, well, should circumstances be very favorable, sometimes to attempt night attacks; at first to employ but a limited number of troops, who, by seeking to become masters of several important points, and to keep near enough to overwhelm the enemy in force at break of day, can derive the greatest advantage from such operations.

The most striking example of a night attack of this kind may be found in the enterprise of the Austrian army against the Prussians at Hochkirch, during the night of the thirteenth of October, 1758. Both armies were in close proximity. Marshal Daun planned the attack with skill, and General Laudon executed it with great vigor. It was favored by the blind confidence of Frederic the Great, who failed to perceive the dangers which menaced him. A spirited attack, executed with several columns, made the Austrians masters of the great battery in the Prussian camp. The battle lasted with great energy until ten o'clock in the morning, when the Prussian army was forced to a retreat, which

---

* As a testimony, see the rash action of the Austrian army at Karausebes, in 1789, under Joseph I. The different columns mistaking each other for the enemy, at night fired upon each other, and more than six thousand men were disabled.—*Note of Author.*

it executed with order and without being pursued, after having lost nearly its whole artillery. The excellent character of the Prussian troops and the prestige of the name of the great captain who had been con-quered, alone saved the army[1], [2].

But if circumstances permitting an undertaking of this nature do but rarely occur, and if the execution of the attack requires great delicacy and much meditation, yet there are other circumstances in which a commander should manifest no hesitation, and where, without running any risk in case of failure, success will be attended with the greatest results.

If troops who are beaten, and upon the retreat, inconsiderately take a position in the evening too near the pursuing enemy, and are unprotected by any material obstacles, the circumstances are exceedingly favorable; then a night attack, executed with but few troops, but led with vigor and intelligence, will always be eminently proper.

In the evening of the Battle of Vauchamps I had the fortune to apply this principle with much success.

On the 14th of February, 1814, after the rash morning's action of Vauchamps, which cost four thousand prisoners to the Prussian army, the enemy commenced his retreat; my corps pursued him with much ardor, and I succeeded in surrounding his rear-guard, composed of a Russian division, with my cavalry, augmented by a reserve of the same arm which Napoleon had placed under my disposition. This Russian infantry bravely resisted the charges directed upon it, and continued to march.

When arrived at Etoges, upon the approach of night, and under shelter of the woods which it had crossed, the Russian infantry arrested its march and made preparations for encamping for the night. I had received an order from Napoleon to halt at Champ-Aubert and take position there; but being perfectly acquainted with the place, having left there only the preceding day, and knowing that the position of Etoges was as bad for the enemy as it was favorable to us, and foreseeing that, on the day following, I would most likely be charged with covering the emperor's movement to gain the vicinity of the corps manœuvring in the basin of the Seine, I considered it of urgency to attempt a sudden assault upon the Russian infantry, and not to wait until they had evacuated the Village of Etoges. I concentrated eight hundred infantry and formed in columns upon the main-road, placing only fifty men upon the right and left into the woods, at a distance of one hundred paces, to guard the flanks, and myself marching with them. I moved forward with the strictest silence, prohibiting the men from firing a single shot, but charging them to fall upon the enemy as soon as they should come sufficiently within his presence. The distance from Champ-Aubert to Etoges is about two miles. In half an hour we had reached the enemy's outposts. The Russian troops, occupied with their camping operations, were scattered about, and only the grand-guards and posts of observation were under arms. With a bayonet-charge we drove everything before us, and threw ourselves upon the village; and in one moment, after having received scarcely five hundred shots, the entire infantry and artillery, consisting of about four thousand men, were in our power, with Prince Ouroussow, the commanding officer.

After a decided reverse and a precipitate retreat, no matter how orderly the latter is executed, we must then be sufficiently distant from the enemy on the evening of the engagement to be beyond his reach; and after an undoubted success we must never hesitate to make a sudden attack during the night upon a beaten enemy, who imprudently has placed himself within reach of the blows of his victor.

I come now to surprises, whose object is to seize a fortified place. Enterprises of this kind have been undertaken several times; some have succeeded, others have failed; and although it is a difficult matter to recognize in any precise manner the circumstances influencing the issue of such undertakings, they may still somewhat be indicated by seeking to discover the conditions upon which success depends.

When such operations can be executed, hesitation should never influence their trial, since their success will oftentimes suddenly change both the system and the character of a war, and secure advantages much greater than those resulting from a gained battle.

Commonly, operations are carried on by means of intelligence established with the inhabitants. Sometimes money is a sufficiently strong inducement to seduce them; but when religious or political passions are exercising their influence, the chance of encountering individuals whose character passes for honorable, and who are disposed to serve us, is pretty often presented. There are likewise enterprises solely executed by the application of ruse, audacity, and courage, which are successful, and whose project is based upon the known weakness of a garrison, or some negligence which has crept into the service. Among the number of the last-mentioned I recall the surprise of Prague by the French army in 1741, which rendered the name of Chevert celebrated, and the taking of Fort Mahon in 1756.

The fundamental principle for success in a surprise, be it favored or not by any one within the place, is to become promptly master of one of the issues leading to the open country. The number of troops, introduced either by stealth or escalade, must always be very limited; and it can never increase as rapidly as the troops uniting for the defence within, nor promptly enough to become formidable to the garrison placed on the defensive; the principal object must then be to send powerful succor with the least possible delay. If this is not done, and the garrison, together with their commander, do not lose their heads, the most hardy enterprises of this kind will always fail.

But it is an important matter to consider that, with even the most favorable elements, failure is yet possible, should the garrison thus surprised be animated by an excellent spirit, and if the soldiers are endowed with great energy, which prevents them from thinking, at the first moment, of the disproportion of their forces and their immediate danger, and which would lead them to think of defending themselves rather than to seek safety in retreat. In that case every soldier fights where he happens to be; the smallest squads, however feeble, band together and defend themselves from behind the door of a house, at the corner of a street, or under the shelter of a wagon, etc. The enemy's combinations are suddenly deranged, and the march of the first troops is checked; it is the first step toward securing the safety of the place. New chances thereafter are presented with every minute; other troops form in the same manner as the preceding, and the garrison is soon rallied and placed beyond any danger, through the power of that moral action which foresees, unites, and, acting by combined movements, comes off victoriously from a strife which at first appeared to be fraught with its destruction.

In like circumstances, the first soldiers who find themselves thus confronted by the enemy should have but one thought—that of the safety of all, and of the glory which always accompanies some great act of devotion.

This sentiment has never been more forcibly illustrated, or has been executed with greater success, than during the surprise of Cremona, on the 1st of February, 1702. The character of the French soldier has never been shown in a more glorious manner; it stands alone in history, and is the greatest exhibition of what courage and valor are able to achieve. Cremona was occupied by the head-quarters of the army and a garrison of eight thousand men. The great extent of the place, the negligent manner with which the interior service was performed, and the security reigning everywhere within its enclosure, as well as the habitual neglect of military duties, were remarked by Prince Eugene of Savoy, and gave origin to the plan of seizing the place by surprise and of making the garrison prisoners. The discovery of an ancient aqueduct, long since abandoned, furthermore favored the enterprise. A priest was gained over, who, with several other inhabitants, made preparations to favor the undertaking. Four hundred grenadiers were introduced in disguise and remained concealed in a church, and some other troops penetrated through the aqueduct. A walled door was demolished during the night; six thousand picked troops, at the head of whom was Prince Eugene himself, the first general of that epoch, apparently took possession of the city; finally, the enemy arrived upon the *place d'armes*, the principal square of the town, and occupied the principal communications before any alarm had startled the garrison. But at the cry *" The enemy is in the town !"* every one woke up and rushed to arms, and small engagements were at once commenced at all points. Marshal Villeroi was taken a prisoner; all the generals except two were either killed, wounded, or captured, and the management of the defence was entirely left to the instinct of the soldiers.   Voices, appearing almost providential, made themselves heard, indicating the movements and combinations through which the garrison could alone be saved; and these troops, surprised in their beds, undressed, and without any officers to lead them, who were vainly endeavoring to join their men, fought with the greatest fury in the midst of this chaos for twelve hours, without either taking food or drink, and in the very middle of winter, devoid of any clothing.   They finally drove the enemy who had at first attacked them from the town, after having placed them in imminent danger of being captured.   And thus this enemy, commanded by an illustrious captain, was entirely mistaken in its calculations, which was that they would encounter only a battalion going to its daily drill, and in the expectation of a reinforcement of four thousand men whose sole duty it was to prevent the flight and escape of the garrison.

Nothing more sublime can be conceived.   If, under such grievous and extraordinary circumstances, a garrison knew how to find safety in its own energy, we may easily judge of what must happen if a garrison does not abandon its trust at the first sight of danger, and endeavors to resist a feeble detachment which has penetrated by means of a surprise; and when the disproportion of numbers is so great between the attacking and defending forces, the mere resistance for one hour will decide the event, since it dispels the effects of being taken unawares, always so powerful; then we arrive within the precincts of reality, a thousand times less formidable, whatever degree of danger it may be fraught with, than those which imagination suggests.

In our days an event like the one I have just related has made our arms illustrious.   It is not generally known, but it is well to recall it to memory, and to transmit the circumstances to posterity.

When, in 1814, the scenes of war were removed from the banks of the Rhine, and Holland evacuated, it became at once hostile to France. English troops, under command of General Graham, were soon disembarked there, in order to sustain the public spirit, and to give some degree of consistency to the approaching revolution.

General Molitor, when evacuating Holland, left garrisons in the most important fortresses; but the state of our armies then did not permit the assignment of many troops to this object, and most likely but depôt-battalions were employed for it.   The garrison of Berg-op-Zoom,

in view of the importance and the extent of that place, was brought up
to four thousand men.   The conscripts, not coming from ancient France,
having deserted, it was reduced to at least three thousand men, and
with this weak body alone the brilliant deed of arms which I am
about to relate was accomplished—an action as glorious to that handful of brave men as to General Bizanet, their commanding officer, attended as it was with the wisest and most far-seeing measures of the
latter, succeeded by still more energetic ones when the moment of action had come.   This case differed from the surprise of Cremona, where
the successful issue had solely been brought about by the obstinate
courage of the soldiers.   At Berg-op-Zoom the soldiers showed themselves eminently brave as well as resolute and energetic ; still it was
especially through their submission to the laws of discipline, and their
ready obedience to the voice of their chief, that they triumphed over
the enemy.

The insufficiency of the garrison had determined General Bizanet to
concentrate all of his troops in the city, and to evacuate the interior
works, among which his small force almost disappeared.   He remedied
the difficulties resulting from this measure by increased surveillance
and a large number of patroles, in doubling his interior posts, and establishing numerous night-piquets, ready at any time to spring to
arms.

General Graham, who commanded the English in Holland, and who
was at a short distance from Berg-op-Zoom, informed of the small number of defenders within the fortress, believed he would be able to seize
it by a *coup-de-main*.   He counted, likewise, upon the *concours* of the
inhabitants, having means of receiving intelligence from within the
place.   He devoted to this enterprise four thousand eight hundred
men, and chose for its execution the night of the 8th of March, the
anniversary of the birth of the Prince of Orange.

The assailant divided his troops into four columns, destined to make
four simultaneous attacks ; the two first were to escalade the rampart,
one to effect an entrance between the gate of Anvers and the seaport,
and the other between the gates of Anvers and Breda ; the third column
was to show itself in front of the gate of Strenberg, in order to make a
false attack ; lastly, the fourth column was to effect an entrance between the city and the port by profiting from the low water.

At ten o'clock at night the third column surprised the outpost near
the Strenberg gate, but was arrested in a complete manner by the fire
of the troops stationed within a stockade of circular palisades which
had been erected for the defence of the fixed bridge.

The garrison at once rushed to arms.

At the same time the fourth column entered between the port without
being perceived by the guard-boat, and penetrated into the city.   But

some troops being sent against it, the column was divided—one portion being held in check and arrested, and the other penetrating through the rampart, where it was pursued.

The second column had been successful in its escalade, and was marching upon the Anvers gate to open it to General Graham, who was waiting upon the glacis with the remainder of his troops and his cavalry. But a strong support of piquets being sent in great haste by General Bizanet to the Anvers gate, it succeeded in preventing the English from seizing it, and the first column having been repulsed in its escalade, was driven away with great loss.

These different engagements, in all directions, occupied nearly the whole night.

At the first dawn of day General Bizanet made an attack with the remainder of his troops, and drove the enemy through the sea-gate and overthrew him. Unable to make a retreat, and torn by the canister-shot of the exterior works, the English columns were thus obliged to lay down their arms, with a loss of twelve hundred dead and six hundred wounded, among whom were two general officers, and two thousand one hundred and seventy-five prisoners, including one general and four colonels, four thousand small-arms, four flags, and a large amount of ammunition, etc.

General Graham solicited an armistice of three days to inter his dead, remove the wounded, and receive the paroled prisoners.

Any comment, after the mere recital of such an action, is unnecessary.

We see from this that French blood has been glorious during all epochs; our customs bestowing extraordinary importance upon the acquirement of military glory, and a proper acknowledgment of the value of the sacrifice of life (a sacrifice which is only adequately compensated when a nation knows how to appreciate the same), has greatly contributed in France to develop the virtue of devotion to the country, the only safeguard for the preservation and the power of nations. The spirit of the army will not change, if only our appreciation of its virtues remain as it is; and may Providence, for the destiny of the country, decree that it be thus; and that those unfeeling minds, who see social happiness in nothing but material prosperity, and whose fatal aberrations show a total unacquaintance with the knowledge of the human heart, may never exercise a power and influence in the councils of the country which, if followed, could not but prove disastrous.

I endeavor in vain to seek for an instance, in my memory, where French troops have been successfully surprised; but there are many to be cited among foreign troops. Two of them happened with the Prussians in the Seven Years' War: at Glatz in 1760, and at Schweidnitz in 1761.

General Laudon had for some time been in correspondence with several officers, garrisoned within the fortress of Glatz, by means of monks who were within the city. The Austrians had scarcely arrived before the place when they opened trenches, and, informed of the time when the officers in their favor would be on guard in the advanced fort, named De la Grue—a fort hewn in rock, and apparently impregnable—they directed a lively attack upon that point; the besieged fled, and the Austrians pursuing with vigor, both parties entered the fortress pell-mell. The troops within reach following the example of those that had preceded them, the Austrians made themselves masters of the place without having experienced the least resistance. In the case of the fortress of Schweidnitz, the surprise happened in the following manner: Five hundred prisoners of war were held in its precincts, and among them an Italian major named Roca, a partisan officer. This officer acquired the good-will of the commandant, and obtained permission to walk at liberty within the fortress. He soon obtained a knowledge of the position of the different posts and the details of the service. Intriguing in the city, he managed to corrupt those who might be of service to him. Upon his reports General Laudon conceived the project of surprising the place, which was executed on the night of the 30th of September. He distributed twenty battalions for four attacks. The commandant of Schweidnitz was at a ball; but on account of several alarming indications he had ordered the garrison under arms, without, however, taking the precaution of sending some person outside of the fortress to ascertain whether the enemy was approaching—so that the Austrians advanced as far as the palisades without having been perceived, and surprised the Stricgauet gate; in the confusion, the prisoners of war, having laid aside their masks, seized the interior gate, and in less than an hour the city was taken and the garrison made prisoners of war.

I will mention, in addition, two surprises which were attempted in our own times, but which failed solely on account of their faulty execution.

In 1796, when the French were placed in position to besiege Mantua, it was considered as possible and very opportune to carry, in the very first night, the T work by surprise. This unlined work protects a long curtain separated from the body of the place, which is only flanked by two heavy towers; it made, therefore, at that time, and did so until the construction of the Fort of Pictoli, which was built by us, the best part of the defence of Mantua from that side. The garrison was considered to be weak and exhausted by illness; three hundred soldiers had been clad in the uniform of one of the regiments in garrison at the place, and placed under the orders of an Italian officer, a deserter from the Austrian army, serving in our ranks; he was to feign a defence of the island

17

upon which the fort is situated, and to appear as being heavily pressed by French troops; to throw himself upon the barrier of the covered way, as if seeking shelter; to have it opened to him, and thus to secure and seize the entrance to the fort. But the officer of whom I have just spoken, not caring to fall into the hands of the Austrians and be hanged, was too timorous in the execution of his rôle, while Murat, who commanded the troops who were to support him, was too slow and circumspect. The troops of the garrison, whose vigilance was excited by reason of so much slowness, were not to be duped by any such farce, which could only have succeeded by means of great activity and extraordinary swiftness.

The second is the enterprise against the Fort of Bard in 1800. Its garrison consisted of barely one hundred and fifty men; the assault would infallibly have succeeded had it been conducted with the least degree of discernment. Colonel Dufour, a brave soldier, but wholly devoid of intelligence and incapable of reflection, was charged with the command of the column which was to carry the gate. Instead of approaching in silence, and placing his ladders against the wall noiselessly, which he would have been able to scale in an instant, he ordered, like an insane man, the beating of the charge when debouching from the village. The garrison, thus advised of its danger, placed itself in a position for defence. Dufour received a ball in the breast, and the attack was repulsed with a considerable loss of men. This check made that hardy and unexampled enterprise necessary—to conduct the artillery by hand, under the walls of that same fort, in the night time, despite the enemy's fire, and thus to clear the defile.*

NOTES.

1. **Battle of Hochkirch,** fought on the 14th of October, 1758, between the Prussians under Frederic II and the Austrians under Field-Marshal Daun. Strength of armies: Prussians—51 battalions, 29,000 men; 108 squadrons, 13,000 —total, 42,000 men. Austrians—116 battalions, 69,000 men; 128 squadrons, 15,000 men; light-horse, 6,000—total, 90,000 men. Loss of Prussians: 246 officers, 8,851 men, 101 cannon, 30 colors. Loss of Austrians: 325 officers, 5,614 men, according to their own statements.

King Frederic, misled through inaccurate reports, was under the impression

---

* I am entitled to claim for myself the merit, conception, and execution of this audacious enterprise, of which I directed all details personally.

The First Consul was only concerned in it as far as he authorized its execution; but justice requires of me to associate with it the name of my chief of staff, then Lieutenant-Colonel De Sénarmont, an officer of great merit and bravery, who afterward rose to be lieutenant-general, and was killed in front of Cadiz. His assistance contributed powerfully to its success. This officer was one of the best known in the artillery service, in which his father before him had already been distinguished.—*Note of Author.*

that Field-Marshal Daun was under full retreat into Bohemia. He immediately resolved to follow him with the bulk of his forces, and left Bautzen on the 10th of October for Hochkirch, a little village some seven miles to the east of the former town, and some three miles to the south of the main branch of the great military high-road from Saxony into Silesia. But his army had no sooner reached the defile of Jenkwitz when it was met by the advanced light troops of Laudon's corps. Soon there was no doubt that the Austrians were assembled in force in his front; and Frederic, repulsing the enemy toward Hochkirch, secured that strong position, leaning his right upon it, and resolving to attack the enemy without delay, despite the objections of several of his generals.

At the same time orders were despatched to General Retzow, who held the left of the Prussian position, to occupy an elevation called the Stromberg, a point of the utmost importance, controlling, as it did, the entire field of battle. This General Retzow neglected to do before the Austrians; and to this failure the disastrous issue of the battle must mainly be ascribed.

The attack was to have taken place on the part of the Prussians; but when Retzow had failed to secure the key of the field, and it had been promptly seized by the Austrians, Frederic was eminently conscious of the danger of his position, and he replied to Marshal Keith's remark: "If Daun leaves us in this camp, he ought to be hanged," laughingly: "It is then to be hoped that he is more afraid of us than of the rope;" but he resolved, as soon as the provisions should arrive from Saxony, to take a different position, and to turn the right flank of the Austrians by a night movement, in order to menace their line of advance into Silesia. For this enterprise the night of the 14th of October had been fixed. The result of a whole campaign, and the lives of ten thousand brave men, hung thus, as the sequel demonstrated, upon the difference of one day.

While Frederic was thus lying in fancied security, almost within cannon-shot of the rapidly-accumulating and threatening masses of his adversary, Marshal Daun had meditated and was resolved upon a brilliant manœuvre, destined to crush the Prussian hero. For the first time in the Austrian general's career against the great king, he had resolved to attack him.

The movement was to be executed by the Austrian left against the strongest position of the Prussians, the wood encircled and fortified Village of Hochkirch, during the night of the 13th of October. At five o'clock in the morning of the 14th the dispositions for the complete turning of the Prussian right were to have been completed, and a strong attack with overwhelming forces was there to be made. A feint attack was to be executed upon the centre, while the Austrian right was to retain its strong position, and only advance to the attack when the complete success upon the Prussian right at Hochkirch was secured.

Unfortunately for the Prussians, the great turning manœuvre of the Austrians was a complete success. Despite the rigor of the climate in those latitudes at that season, the dense fog which had settled over the whole country, and the immense difficulties of marching through pathless woods under such circumstances, this march was splendidly executed, and reflects the highest credit upon the Austrian arms. To completely deceive the Prussians, during the 13th of October, and continuing during the night, large bodies of troops were assiduously felling trees and constructing abattis in front of the Prussian position, persistently challenging each other, and keeping up a great noise, as if from that quarter the main danger to the Austrians threatened.

In front of Hochkirch was then a little birch wood, before which the dark masses of the Austrians had assembled within musket-range of the Prussian advanced posts, at four o'clock on the cold, foggy morning of the 14th of October. So great was the security of the Prussians that the infantry lay, undressed, in slumber within their tents, and even the cavalry, save a few regiments, among which that of Ziethen, had unsaddled their horses.

When the village-clock of Hochkirch struck five o'clock, the Prussian free-battalions, which held the birch wood, were attacked by the infantry of one of Daun's divisions and the Pandours of Laudon's corps, and, after a short struggle, thrown back upon the three battalions of the extreme Prussian right, which hastened to the succor of the retreating advanced guard. Formed with the alacrity of veterans, they opposed to the Austrians an impenetrable front, and had already driven them beyond the birch wood, when they were suddenly assailed in rear by Croats and infantry who had penetrated to their abandoned camps. Still they continued the struggle, but were at last compelled to cut their way through the rapidly-augmenting masses in their rear, and to retreat to the village, behind the shelter of a battery of eighteen heavy 12-pounders.

Simultaneously with this attack, Laudon had repulsed the Prussian videttes of the right flank, and General Ziethen was compelled to withdraw behind Hochkirch likewise. Quickly taking advantage of his easy success, Laudon established a battery of eight guns upon the heights of Meschwitz, from which the whole camps of the right wing were at his mercy.

Now thoroughly alive to the dangers of the situation, the entire Prussian lines were under arms, and three fresh battalions of the second line advanced to the succor of the troops who had been driven from the wood of birches, where, after having retaken it, they were a second time forced to give way to overwhelming numbers. Upon this success Daun advanced his whole line and took the heavy Prussian battery in front of Hochkirch, throwing the Prussians entirely into the village, and occupying the whole ground in front and flank, with the exception of a few gardens and the church-yard—the latter defended valiantly by the battalion of Major Lánge. It was now half-past five o'clock.

Marshal Keith and General Ziethen made several fruitless attempts to retake the heavy battery, with invincible resolution. The former, who so long had illustrated the Prussian arms, here fell, and Ziethen was disastrously driven back to Hochkirch.

The village itself now became the theatre of one of the most bloody episodes of modern times. Up to this time Frederic had never doubted that the demonstration of the Austrians was but a feint. But now he ordered heavy reinforcements for the possession of Hochkirch. Prince Francis of Brunswick and Prince Maurice of Dessau led their troops against the hosts of the enemy. The former, while endeavoring to retake the great battery, was violently assailed in both flanks; his troops, after a stout resistance, were overpowered, and, after the fall of their heroic commander, retreated in utter disorder behind the village. Hochkirch was in flames, and the Prussians were rapidly giving way. But another band of heroes, under Prince Maurice, issued from the disorder, and, driving the Austrians beyond the village, retook the birch wood; and, had he received reinforcements, he would have held himself there. He was at last obliged to give way, and with the utmost difficulty he effected his retreat. Amid all the storm raging around him, Major Lange, with his weak battalion, had maintained his important position within the enclosures of the church-yard of Hochkirch. The successive attacks of seven regiments he had withstood; but when Prince Maurice gave way the flood overpowered him also. His fate was a sad one. After prodigies of valor his brave command was entirely destroyed, and this intrepid officer himself fell with the band of heroes he had led

Ziethen, with his renowned horsemen, had meantime charged the Austrian cavalry repeatedly, but without effect, and was at last driven beyond Hochkirch with great loss.

From the wreck of the Prussian right wing Prince Maurice of Dessau made a last attempt with four battalions to retake the ill-fated village, now burning furiously. After a severe contest the forlorn-hope gave way; the heroic prince fell dead, pierced by two musket-balls.

At this crisis the king placed himself in person before three regiments and one battalion of the guards. Incredible efforts they made to wrench the victory from the hands of the Austrians; they even penetrated to the right of the village to the very spot where their right flank had sustained at first the conflict. Their attack might have succeeded had not, at the moment they charged, a repeated rear-attack thrown them into disorder. With this last attempt ended the efforts of the Prussians to regain the village, which henceforth remained in the enemy's possession.

It was now after seven o'clock in the morning, and the heavy fog at last cleared away and revealed to the king the extent of the disaster. The plain to the right and rear of Hochkirch, toward the defile of Dresa, was dotted with the retreating and advancing columns, and shattered remains of both contending hosts in confusion. Both generals appear, at this moment, to have understood the situation alike. Frederic, now conscious of the great disaster which had befallen him, rapidly sent orders to establish a new line of battle to the rear of the former, and by an entire change of front, which brought the Prussian right wing nearer toward the high-road to Bautzen. Marshal Daun, on his part, rapidly concentrating his scattered columns, endeavored to cut off the retreat of the Prussians upon Saxony, and had nigh succeeded when Major Moellendorf repulsed his columns with a hastily-collected body of infantry and artillery, and thus achieved the great glory of keeping open to the beaten but not discouraged Prussians their last and only line of retreat. Simultaneously with this attempt, General Colloredo was repulsed on the other flank while trying to turn the left of the new Prussian position, and under both trials the Prussian army finally and heroically formed their new line of battle, confidently awaiting the further attacks of the enemy.

Marshal Daun contented himself, seeing the *sang-froid* and resolution of the Prussian right, to await the issue of the Duke of Ahremberg's attack upon the Prussian left, from which he expected to give the finishing stroke to the fortunes of his adversary. During the sanguinary struggle on the right, the Prussian general on the left had been permitted to remain unattacked, conformable to the orders of Daun. But when the successive reinforcements, drawn from the centre toward Hochkirch, threatened to weaken the line so as to endanger his safety, he slowly retired, closely followed by the Austrian overwhelming forces, to the shelter of a heavy battery of thirty guns, yet intact. Finally, at the moment when Frederic was endeavoring to safely withdraw his right, the left was fiercely assailed by Ahremberg, who, with surprising rapidity, overthrew all the forces in his front, charged, and took the great Prussian battery of the left, and made many prisoners. But this precious advantage was neither followed up by the duke nor by the marshal. The latter was satisfied with the honors of Hochkirch, and remained a peaceable spectator of the wise retreat of the king, which now took place from every portion of the field.

General Retzow, who had neglected to seize the Stromberg, without the possession of which Daun would never have thought himself strong enough to attack Frederic, was confronted by the Prince of Durlach, entirely cut off from the main body of

17*

the king's army, and in an extremely perilous position. If the Duke of Ahremberg had failed to reap any fruits from a decisive repulse of the enemy, Prince Durlach failed still more deplorably against General Retzow. The latter eluded him perfectly, and, although making a detour to avoid the enemy's main body, a portion of his command arrived in time upon the field of battle to repulse a charge of the enemy's cavalry upon the flank of the Prussian main body. Retzow himself, with the remainder of his forces, joined later. With this skilful extrication from a dangerous position, if he did not atone for, he at least considerably softened, his previous failure.

To cover his retrograde movement Frederic placed his cavalry in large intervals upon the plain; the baggage passed through the defile of Dresa; the infantry followed, and he took a new position upon the mamelons, called Spitzbergen, to the north of the battle-field, and half-way between Hochkirch and Bautzen—in fact, the same position which the Allies subsequently occupied after the Battle of Lützen in 1813.

2. Celebrated turning manœuvre of Frederic after the battle.—Daun, after the decisive repulse of the king, took a new position in advance of Hochkirch, upon the line of the Spree, facing Saxony, hardly three miles from the king's new camp, and covering effectually the grand route into Silesia—contenting himself to watch the Prussians, and thus forgetting the important maxim in war, that the fruit of a victory depends upon the profit we draw therefrom—the most desirable talent of a general.

The condition of Frederic after the battle would have been a disheartening one to any second-rate general, and would most probably have led to the conclusion of the struggle; but to the king it presented but new chances to improve his situation; and, sustained by the fine spirit of his brave veterans, he resolved, by a grand detour, after having received reinforcements, to fall upon the rear of the Austrians. As soon as he had gained his new camp the resolution was instantly taken and announced in these words: *"Daun has permitted us to escape; the game is not lost; we will repose a few days, and then march upon Silesia to deliver Neiss.*

The great route from Saxony to Silesia, shortly after issuing from Bautzen, divides into two branches, both running nearly due east. The northern branch, after ten miles, passes the Town of Weissenberg, and unites again with the southern branch at the Town of Reichenbach, distant from Bautzen fifteen miles; whence the road continues past Markersdorf, where Duroc fell in 1813 at the side of Napoleon, to the fortress of Gœrlitz, on the Neiss, at a distance of six miles.

The Austrian camp covered both of these roads in advance of Weissenberg, and was some five miles from Bautzen—Frederic's camp being an equal distance to the north of that town. From the Austrian camp the distance to Gœrlitz was about eighteen miles, or one good day's march.

To the north of these roads the country is exceedingly hilly, and dotted with woods and small villages, traversed by little streams, effectually covering any movements, but, nevertheless, dangerous on account of the enemy's numerous light troops scouring the country, and which could only be executed upon miserable cross-roads, almost impassable in the rigorous season which had set in. The king's project was to transport the bulk of his stores and artillery, and to execute a rapid march through this difficult country, so as to gain Gœrlitz before the enemy. This point once reached, there appeared no further obstacle to his rapid advance for the deliverance of the Silesian fortress of Neiss, which, being besieged by the Austrian general,

Harsch, with thirty thousand men, was already believed to be near reduction. Neiss once delivered, he was to return to Saxony upon rapid marches, for the succor of Dresden, only held by a small force.

To deceive the Austrians the sick and wounded were transported north to Glogau. The munitions of war were sent, at first, directly north, but suddenly turning at Kumerau to the right, disappeared in the intricacies of that difficult country. Both movements were well known to the Austrians, who now felt perfectly sure of the intention of the king to evacuate Saxony, and to retreat within his hereditary dominions.

On the 24th of October, at six in the evening, the trains and baggage, under Braun, were sent off, and after a march of six miles, directly north, parked at Neudorf. The infantry occupying the villages in front and on the flanks, retired to camp at ten o'clock at night in the greatest silence, leaving the advanced guards of cavalry there stationed until next morning. At ten o'clock at night the tents were struck and the army began the march by lines and by the left. The vanguard marched to the north-east upon Dresa, passed the Little Spree, changed direction to the right, and marched all night upon Ullersdorf, fourteen miles from camp, a little village whose left flank and front being protected by small watercourses and three lakes, formed an eligible position for defence. The first column, composed of the entire infantry, followed the same route; the second column passed Neudorf, upon a line to the rear of the infantry, and was there followed by Braun's convoy, arriving at the position of Jenkendorf, near Ullersdorf, on the 25th, at noon. Finally, the rear-guard, under Prince Henry, occupied the abandoned position all night, and followed the main body early in the morning. Nearly the entire Prussian army was concentrated, at break of day of the 25th of October, at Ullersdorf, some ten miles to the north of Gœrlitz.

The Austrians were only aware of the king's departure late at night on the 24th, when the king was yet some fifteen miles from Gœrlitz. The distance of their camp from that fortress, upon a splendid military high-road, was but eighteen miles, while Frederic had to execute a march nearly double that distance. Had Daun at once marched, it is probable that he would have gained possession of it before Frederic; but he contented himself to send General Caramelly to pursue the rear-guard, and only in the afternoon of the 25th he detached the grenadiers, the reserve, and two regiments of hussars from Reichenbach to observe the march of the Prussians. Lascy, who commanded this detachment, seeing Frederic encamped at Ullersdorf, no longer was in doubt, and resolved in consequence to set out at night and occupy Gœrlitz and the heights of Landskrone, in advance of that town.

Frederic, however, here again foiled the Austrians. Setting out at night, two o'clock, with the advance guard and the entire cavalry, he marched across the country upon Ober-Rengersdorf, five miles from Gœrlitz, followed at three by the infantry. At break of day he met the Austrians in his way, but overthrowing them, after a sharp combat, he gained the town.

The consequences of this brilliant manœuvre were of the greatest importance. Although closely followed by the Austrians, he rapidly marched into Silesia. The siege of Neiss was abandoned on the 5th of November, and on the 8th already we see Frederic marching upon Saxony to deliver Dresden. Thus, in the middle of winter, Frederic had, in six weeks, twice changed the direction of the war, delivered two places of the first rank, torn from a victorious army all the fruits of victory, and thrown it back upon its own frontiers.

CHAPTER X.

DEFENCE OF FORTRESSES.

Essential conditions—Importance of the office of commander—Glories compared —Of true courage and the truly brave—Defence of Grave by Chamilly—Of Lille by Boufflers—Of Saint-Sébastien by Rey—Of Burgos by Dubreton—Of Wittenberg by Lapoype—Necessity of a legal inquiry in all cases of surrender —Sorties—Their opportuneness and aim—Surprises easy—Prague in 1741— Rodrigo in 1812—Badajoz, Saragossa, Genoa.

Notes.—1. A singular Polish Council of War. 2. Siege of Grave in 1674.

The first element requisite for the successful resistance of a fortified place is a good commandant. If to this first and indispensable requirement a sufficiently strong garrison and a large amount of stores of all kinds, such as provisions, munitions, etc., be added, the most extraordinary results may be obtained..

The fortifications themselves may be more or less perfect, but this always desirable perfection is a matter of little importance when compared with those results owing to the courage and the resolution of him who directs the defence.

The commandant of a place is its very soul; it lives in him and through him. If, at the beginning of a siege, the garrison be a bad one, it will soon become a good one under a good commandant, who knows how to awaken its sentiments of honor, patriotism, and glory, which sometimes grow dull in the soldier's heart.

It is an excellent thing to gain battles; the glory which falls upon the chief is dazzling; success calls forth every sentiment of enthusiasm and admiration; but it is a still more excellent thing, more meritorious at least, to defend a place for a period exceeding certain supposed limits.

The glory of a victorious battle, however brilliant it may be for the general, is necessarily always divided; that which the commandant of a fortress acquires belongs to him almost entirely. That glory is his own work; it is the fruit, not of combined action accomplished under certain circumstances, but of a long series of uninterrupted, persevering efforts, renewed without cessation, with the same consciousness of their inutility, if succor does not arrive in good time; and the efforts of each succeeding day are not rewarded by the prospect of the pleas_ ures of victory; they are, on the. contrary, associated with the ever-present feeling of our relative weakness, and their only aim is to retard the success of, and not to triumph over, the enemy, without the least hope of changing the ultimate issue in any way.

Every man of nerve has always sufficient courage and energy for the requirements of a period of twenty-four hours; and if success attends the efforts every man appears to be a hero! But how rare is it to find the same courage, tenacity, and ardor in reverses; the truly brave alone will then show these qualities, and their number can be easily counted.

But the commandant of a besieged fortress is placed in a still more difficult position; he must not alone, from the very beginning, preserve his moral courage—a virtue so seldom accorded to man by Providence—but that same courage must grow within him in measure as circumstances become more difficult, and when it would be natural to suppose that it should sink—because his deportment must counterbalance the effect of the miseries and sufferings which he, with the garrison, will be obliged to endure. The commander alone appears to be interested in the defence, because he reaps the glory almost entirely, while those under his orders only endure the sufferings. Moreover, whenever a commandant is inclined to surrender, he will always find those who will come forward to applaud such an action, and officers ready to make him easy should his mind still entertain any scruples or doubts as to the propriety of his resolution; and whenever a commander is inclined to submit to the decision of a council of war the question whether it be time or not to capitulate, the result will always be in the affirmative; and it even sometimes happens that there are persons who would not protest against the surrender, did they know that their opinion would be of any power in changing the decision of the majority of the council.[1]

Nothing appears more worthy of admiration than the defence of a place prolonged to the utmost limits, but there is, likewise, nothing which happens less frequently.

Justice demands, then, that we should render the names of those immortal who have acquired glory of this kind.

The most celebrated defence known to history, in modern wars, is that of Grave,[2] on the Meuse, by Chamilly, in 1675; nothing can be compared to it. This city had received the depots of the army from the time of the invasion of Holland by Louis XIV, and enclosed a large amount of stores. Its extent is of mean dimensions, and it was garrisoned by five thousand men, who defended themselves during five months of opened trenches against all the efforts of the Prince of Orange, whose loss amounted to thirty thousand men; and Chamilly only surrendered upon the king's express order, carrying with him to France all of his ordnance.*

---

* I have always had the ambition of being charged with the defence of a large fortress, having the innate consciousness that such a task would not have been beyond my capacity. Had there even been any occasion in my experience, I would have ordered the reprint of the journal of the Siege of Grave, so that every officer, non-commissioned officer, and soldier would have had an example before him, worthy to be followed. Should the day ever come when regiments are provided with libraries, this work, of the greatest interest for the military man, should not fail to find a place in it.—*Note of Author.*

After the last mentioned admirable defence, that of Lille and its citadel must be ranked next. Marshal Boufflers, who commanded there, acquired immortal glory by it.

In our own days sieges have not been of frequent occurrence; still we can not pass that of Saint-Sébastien in silence, which was commanded by General Rey—resulting in a long and obstinate defence, which caused to the English army very considerable losses.

The defence of Burgos, under the orders of General Dubreton, who, though attacked by a less powerful force than he himself disposed of, still acquired some glory; and that of Wittenberg upon the Elbe, under General Lapoype, may likewise be mentioned.

But in comparison with a few extraordinary instances of resistance, which we can not but admire, how many mediocre defences and culpable, unpunished surrenders are there, which are looked upon by public opinion with a degree of indulgence entirely unmerited.

The preservation of a fortified place is a matter of such capital importance, influencing, as it does, sometimes in the most powerful manner, the fate of an army or even of an entire country, that its surrender should always be the occasion for a legal inquiry, in order to bring out, in all their clearness, the whole train of circumstances of the defence which led to the capitulation. The commanding officer should then be either severely punished, or be rewarded and overwhelmed with praises; a different course of proceedings ought not to be admitted.

The navy regulations prescribe the trial of every captain of a vessel who has lost his ship, no matter how the occurrence took place. If he performed his duty, he is acquitted and honorably dismissed.

Any indulgence in legislation in matters concerning the sea will be easily understood, because, upon an element as changing as the water, circumstances much more powerfully influencing and conquering science, vigilance, and courage, may happen. But upon *terra firma* nothing varies; unless the reduction be caused by the want of army stores of any kind, there can be no legitimate excuse; we can but choose between either censure or praise. The military regulations upon this point must be vigorously enforced, and if a commander surrenders before the *enceinte* has a practicable breach, and before he has sustained at least one assault, he has committed a crime, and should receive due punishment.

I shall not enter into technical details as to the attack and defence of places; special works have treated of these matters sufficiently satisfactorily. I will be content with making some reflections upon the general direction to be followed in the defence.

In the case of large fortresses, the custom is too prevalent of making sorties long before the commencement of the siege, and a portion of one's means, forces, and confidence is thereby expended, which to retain for the moment when courage and vigor are of the greatest necessity, would be much more useful and important. By going any distance from the fortifications we lose their support, and voluntarily deprive ourselves of a succor which establishes some sort of equilibrium between the troops of the garrison and the attacking force. I would, therefore, at any rate, at least have some hopes of raising the siege by making a sortie with the larger portion of the garrison, and to remain sufficiently near to have the efficacious support of the cannon of the fortress.

But if these sorties should be prohibited, those whose object it is to destroy works begun by the enemy can not be made too often ; the principal aim being to arrest the enemy and to gain time, which can be attained by giving him frequent causes for alarm and in bringing about short and lively engagements, which will force him to recommence the same works several times. In measure as the enemy approaches the place and the siege is progressing, sorties, with but few troops and upon a restrained field of battle, should become more frequent. Finally, at the very time when the great proximity of the enemy puts so often the idea into the heads of commanders to surrender, the true defence should commence ; and it even appears to me that it ought never to be finished, if with every day new obstacles are prepared, if interior entrenchments were beforehand constructed, and such dispositions made that the besieged would never be completely deprived of the fire of his artillery, but would always retain some well-covered pieces of cannon for the defence of the breach. This precaution alone, which should be particularly considered and taken, may decide the safety of the place during several days, and add much to the glory of the defence.

I will close this chapter with a remark by which commandants of besieged fortresses can not too much profit. They should particularly guard against surprises, because the more a thing appears improbable, the greater will be the effect when it happens. A brave garrison defends a breach, and the enemy will not be able to contend long against it ; but if at the moment when all attention is directed upon the defence of an open point, the defenders hear that the enemy has effected an entrance into the fortress by escalade at another point, then men's minds will be upset, the defence of the breach will be abandoned, and the place is taken.

The strictest surveillance of all points should, therefore, never be relaxed, and those appearing in least danger of being attacked, because they are considered impregnable, should be guarded most—since they are the very points which the enemy will choose in preference, because it is evident that if these points appear to sustain themselves, it is not likely that any one will be charged with their defence.

In 1741 the fortress of Prague was the object of very noisy night attacks upon two points on the part of the French army ; and while these demonstrations attracted the attention of the entire garrison, other troops were silently directed upon a point of the *enceinte* of the new city, some considerable distance from them; they scaled the wall with a single ladder, mounted the rampart, and, finding nobody defending the place, the gate near was at once opened and the city, with its garrison, taken almost without any engagement.

In our days, in 1812, the garrison of Rodrigo bravely defended a practicable breach in the body of the place and repulsed the enemy; but fifty English soldiers scaled the castle with ladders, which commanded all other points, having a revetted scarp, and being very high; they created alarm within, disorder ensued, and they thus became masters of the town.

The Town of Badajoz was besieged in the same year, and well garrisoned, commanded by a distinguished soldier, General Philippon, who already, the year preceding, had sustained a glorious siege. Entrenched around the breach, he repulsed the assaults of the enemy; but the castle, the walls of which were eighty feet high, were scaled by fifty men, alarm and disorder spread through the town, and it was taken.

Under no pretext should watchfulness ever be relaxed. Means of resistance must be everywhere created; and even in those parts of the fortress which appear least exposed to an attack, especially at the commencement of a siege, and when the enemy may and should suppose that all means of defence have been concentrated upon the points upon which he is about to direct his attacks.*

NOTES.

1. **A singular Polish Council of War.**—In 1674, when the Grand-Vizier, in his retreat after the defeat near Leopol, in Gallicia, invested Trembowla, a small town, strongly fortified, in the province of Podolia, and summoned the commandant of the Polish garrison, Samuel Chrasanowski, a Jew, to surrender, he received the following reply: "Thou art mistaken if thou expectest to find any gold within these walls; we have nothing here but steel and soldiers. Our number, indeed, is small, but our courage is great."

The Turks at once attacked the town upon all points, with an overwhelming force and great vigor. A council of war, convoked by the commandant, advised to open negotiations for the surrender of the town, in order to save it from the fury of the Moslem hordes; when suddenly the wife of the commandant, a Jew-

---

* I have not included among the number of remarkable defences that of Saragossa by the Spaniards, because it belongs to another class of events. An immense population, refugees from the country, with a large amount of provisions—a population excited to fanaticism by religion and patriotism, whose numbers being constantly double that of the besiegers, and whose daily losses were hardly felt, occupying those immense and indestructible convents as strong as real fortresses—could and ought to have for a long time frustrated our efforts. But circumstances of this kind can not recur, and this defence can offer no precepts which might be of use in a regular war. As to the Siege of Genoa, it was a remarkable and great operation, but rather the defence of an entrenched camp, and not that of a fortress. —*Note of Author.*

ess, appeared in their midst, and, with two daggers in her hands, thus confronted her husband, saying: "One of these daggers is destined for thee, if thou surrenderest this town; the other I will plunge in my own breast."

After this impassioned appeal the council, of course, had nothing more to do but to submit, and dispersed; and, fired by the resolution of this Jewish heroine, a vigorous resistance was continued, in which the brave wife of the commandant, together with others of her sex, participated by supplying ammunition to the soldiers and by stimulating their courage.

By a singular providential coincidence this noble devotion was not destined to remain unrewarded. King Sobiewski, at the head of a Polish army, a few hours afterward appeared to relieve the garrison, completely routed the Turks, and upon their retreat took eight thousand prisoners.

(Mentioned in the Polish histories of Naraszewiez, Bantkie, and Lelewel.)

2. Siege of Grave in 1674.—The celebrated journal of this siege, alluded to by Marshal Marmont, would be too voluminous for this work. The following account is therefore given in its place:

Rabenhaupt, the Dutch general, was still in front of Grave, which had been invested for a long time, and was being besieged regularly. This place, built upon the banks of the Meuse, was strong by its position. The French, anxious to keep it while abandoning other forts, had augmented its fortifications, and mounted more than three hundred pieces of cannon. A garrison of four thousand men, under command of Count de Chamilly, defended it. De Betonce, Saint-Just, and Saint-Louis, old officers, and the Marquis of Guiscard, commanded the regiment Normandy, sustained by those of Bourgogne, Languedoc, Vendôme, and Dampierre.

Rabenhaupt had, since the month of July, made himself master of the posts likely to facilitate the enemy's communications. By means of friends within they made an attempt to retake Fort Ravesteyn, but the traitors were discovered and punished. The Meuse only remained to the besieged to pass their soldiers across to forage on the other bank. Several engagements took place with the horsemen, who attempted to pass the river in order to destroy a dyke which was in their way. Colonel Hundebek bravely defended it. Four hundred cannon-shots were fired upon him in a single day, doing, however, little harm, the balls passing above him. The entire cavalry within the town, sustained by four hundred and fifty infantry, made a sortie to carry the guard off, stationed in the direction of Velp, but met such a vigorous resistance that they were obliged to retire. Nine French officers, who had dined with Count de Chamilly, desirous of signalizing their debauchery by something extraordinary, advanced as far as the dyke with twenty soldiers. Animated by the fumes of wine, they at first overthrew the opposing force; but Colonel Hundebek having arrived to the succor of his retreating men, the attacking party took refuge within the church of Velp; but seeing that the building was about to be set on fire, they were not foolhardy enough to remain in it, preferring rather to surrender than to be burned alive. Another still more audacious performance took place. Three hundred horsemen, with as many footmen mounted behind them, in full daylight, made a sortie toward Velp, overthrew the company of Rainmaker, there on guard, took the greater part of the officers prisoners, and would have caused much more disorder had not General Spaen obliged them to retreat into the city. Rabenhaupt, who had his head-quarters at Balgoyen, pushed the attack, threw shells, red-hot balls, and fire-pots into the city. He erected a battery with which he destroyed the steeple and a number of houses; ordered Goldsteyn to take a position, with five

18

regiments and five field-pieces, at Overasselt, to harass the besieged from that quarter; and Colonel Winbergen to pass the Meuse with eight hundred men, and to encamp within musket-range of the town. Count de Chamilly ordered a sortie with two hundred and fifty footmen to drive those near Colonel Hundebek's quarters from the trenches. DuPas, who had been Governor of Narde, took part in this action, and sought, by extraordinary bravery, to efface the disgrace of his condemnation.* He charged with such courage that he drove away those who defended the approaches; but Colonel Litzau, on guard near head-quarters, came promptly to their succor; the French were in their turn repulsed, and DuPas left there a life which had become insupportable to him, and which he sold dearly. The besiegers battered with such success the ravelin on his side of the Meuse, that Count de Chamilly, foreseeing an assault which his men would not be able to sustain, ordered them to retire after having sprung the mine, and to entice the Hollanders; but the latter took the precaution to wait until it had exploded, and only six or seven men perished. Upon this ravelin a battery to bombard the platform upon the bank of the Meuse was erected. The besieged, fearing that their gun upon this platform might fall into the river, withdrew it, which rendered the position useless, and gave greater scope to the besiegers. Three rows of palisades would have to be forced before they could become masters of the entrenchments. It was done; but the besieged, having sent succor promptly forward, regained and alternately lost it three times. Finally, when the Hollanders pressed the regiment Bourgogne heavily, which held the position, it retired, springing a mine which killed several persons and threw the remaining troops into confusion. Chamilly, profiting from this disorder, ordered the regiment Normandy to advance, which drove the Hollanders from every position they had occupied. The engagement lasted until nine in the evening. The following day a suspension of arms was agreed upon, to inter the large number of the dead. The horses had been killed, as forage was scarce in the town; still there were plenty of provisions left for the men.

While Grave was thus invested from all sides, Chamilly did not cease to receive news from without, by means of swimmers who passed the Meuse below the water, but even money to pay and animate his garrison, which began to fear that the siege, which had lasted too long already, would be prolonged indefinitely. They were in this inquietude when the Prince of Orange arrived at camp with a succor of more than ten thousand men. (October 9.) The forces before Grave amounted now to more than twenty thousand. The prince reanimated the dis-

---

* DuPas had been appointed Governor of Narde upon the recommendation of Turenne, in the preceding year; a place badly provisioned and fortified, with a garrison of three thousand men, and but eighteen guns, some of which were not mounted.

The Prince of Orange shortly afterward invested it with twenty-five thousand men, and four days after the trenches had been opened DuPas capitulated.

Louis XIV, irritated at this feeble defence, disbanded all the troops engaged in it, and broke the officers.

The governor was treated much more severely. He was placed before a council of war; condemned to be dishonored; the hangman publicly broke his sword upon his back; and he was sent a prisoner into Grave, where Count Chamilly, enraged at the treatment this brave man had received, offered to return him his sword. This DuPas declined, until he should have regained his honor by bravery.

heartened soldiers by his presence; the bombardment was redoubled, and as many as two thousand shots were counted on both sides in two days. Three Dutch regiments attacked the entrenchments, which the besieged had constructed upon the dyke outside of the Bruges gate; the regiment Vendôme, which guarded them, at first took to flight; but Count de Chamilly, sword in hand, forced them to return to the combat, and supported them by the regiments Languedoc and Dampierre, posted near them expressly for that purpose. However, fearing their being forced, he ordered the mines to be fired, which, springing too soon, nearly buried as many of his own men as of the Dutch. Lastly, to preserve this post, he advanced the greater portion of his troops, who obliged the assailants to abandon their enterprise, after having lost some good officers, among whom the nephew of Rabenhaupt. On the part of the besieged, Count Guiscard was wounded while visiting the parapet of the covered-way, but the wound was slight.

During the night of the 13th of October the prince ordered the attack of the counterscarp simultaneously upon the Bruges gate, the dyke of Ravesteyn, and Bastion Prince Maurice; when a fierce combat ensued, until two Dutch squadrons, having passed the bridges across the moat, formed in battle order upon the glacis. The French, surprised to see themselves attacked in that quarter, abandoned the counterscarp; and two captains, with forty-eight soldiers, having passed the palisades, penetrated to the covered-way. De Chamilly, apprehensive of their further advance, detached eight troopers from each company, who, however, were obliged to retire after a single discharge. Du Fail, a captain of cavalry, having been wounded in the thumb at the commencement of the action, went to the town to have his wound bandaged, and the surgeon not being able to attend to him as promptly as he desired, he had his thumb cut off, and returned to the fight with intrepid courage. The regiment Normandy, sent to dislodge the besiegers from the covered-way, engaged them so vigorously that they were obliged to beat a retreat; and, meeting the same degree of resistance upon all other points, they abandoned the counterscarp, leaving behind several prisoners and many dead. The next day an attack was made in full daylight upon the gate of Bruges, where the regiment Languedoc resisted so bravely that they abandoned the attempt, after a furious combat. Colonel Goldsteyn, undismayed, returned to the charge, carried the entrenchments on that side, and established himself upon the glacis, while Hundebek made himself master of those of Ravesteyn; then, having carried the moat in front of the covered-way, he effected a lodgment there. Forty-five troopers repulsed them three times, but their number being reduced to fifteen, they were obliged to retreat to the city and permit the besiegers to perfect their lodgment. Early on the morning of the 16th of October they advanced from the same side to the Place d'Armes. There two pieces of cannon of the besieged killed many men; and when they were about being stormed, the French sprang a mine, which buried a portion of the assailants. One of their captains was found half-buried without having received any injury, and taken to the city. The others returned to charge the French, gained and lost four successive times the entrenchments upon the glacis, but were finally driven back to the approaches. It would be difficult to cite a siege more bloody, or a place better defended. No day passed without several engagements. The besiegers were often repulsed, but the works progressed, without intermission, for a general assault. On the 25th, the besiegers having sprung a mine which they discovered in their sapping operations, a quantity of grenades, which the besieged had near the gate of Bruges, exploded, and made a great breach in the covered-way.

As the number of dead and wounded was very large, a suspension of arms was agreed upon, during which Count de Chamilly offered to surrender the place if an honorable capitulation should be accorded to him. Some time before he had sent one of his swimmers to Maseik, who returned with an order from the king to surrender the place to the Prince of Orange. The besieged were short of provisions; the soldiers reduced to drink water and to eat half-rations. The garrison was fatigued and reduced, and the place so breached that it could have held out but a few days longer. One of the principal articles of capitulation was to permit Chamilly to carry off, with the French arms, twenty-four pieces of cannon.

Thus Count de Chamilly surrendered Grave, by the express order of the king, to the Prince of Orange, on the 26th of October, 1674, after a siege of ninety-three days.

PART FOURTH.

PHILOSOPHY OF WAR.

CHAPTER I.

MORALS OF SOLDIERS, AND HOW TO FORM THEM. FORMER ARMIES AND THOSE OF THE PRESENT TIME.[1]

Elements of valor—Means of development—Discipline—Corporeal punishment—Inadmissible, in which armies—Power of public opinion in French army—Praise and censure; emulation—Dignity of man respected when punishing a soldier—Contempt shown only to a coward—In what consists valor—Different gradations—Severity—Confidence—Power easily exercised in peace—Difficult in danger—Confidence gives energy—Duty of chief toward soldier—Sublime impulses of the heart—Two facts—Activity—Health—Military exercises—Instruction; utility of great manœuvres—Emulation—Public works—Association of soldiers with them—Egypt, Holland, Dalmatia—Names of regiments, etc., inscribed upon public works—Great camps of instruction—Their importance—Their effect upon the army of 1805—Must be made permanent—Countries without cultivation—Three months' exercises—Algeria—Regular armies of Europe—Feudal state—Sovereigns dependent upon vassals—Finances—Free companies—Regular revenues—Order and economy—Armies means of civilization—Obstacles created by lords—Voluntary enlistment—Disorder favorable to them—Profession of soldier a resource—Often results in large fortunes—Visconti, Sforza, Scaligeri, Norman adventurers—They became agents—Regiments, a speculation—Ferdinand II—Wallenstein—Regiment, a property—Origin of proprietary—Recruiting—Voluntary enrolment insufficient—Why—Great effect upon civilization—Blood-tax—Salutary influence—Comparative morality of armies composed of citizens' sons and vagabonds—Two systems of distributing men in armies—Usage in France and Russia—The best—Council of war of 1828.

Notes.—1. Success in organizing the Confederate army. General Samuel Cooper, Confederate army. Confederate soldiers and United States soldiers. 2. Duke Wallenstein.

Three things are essential to impart valor to troops: love of order, habit of obedience, and confidence in one's self and others. Such are, in their moral aspect, the principles upon which the structure of an army reposes. Bereft of these, an assemblage of men has no consistence whatever; it will realize not even the slightest expectations, and is sufficient for no possible requirement.

18*

Nothing, therefore, should be neglected to develop these three principles in the soldier's heart, and to unite to the morals of military men those customs which I shall designate by the name of military virtues.

It is requisite that discipline, which is submission to regulations and to the will of the legal chief, be observed without any relaxation; and that each one, in the rank which he holds in the military organization, be unceasingly mindful of the fact that he can only command his subordinates by virtue of the title which his own obedience to his superiors confers upon him.

Although discipline must always be severe, when a grave failure has been committed, it should, nevertheless, have its gradations in its different applications.

In countries where elevation of sentiment, delicacy of manners, and dignity of character forbid the application of corporeal punishments, it is important to unite the influence of moral power as much as possible with punishments.

The French army, in particular, has always offered to an intelligent chief frequent opportunities to make this resource useful. Praise and censure, distributed in the proper manner, and the application of the talent to excite a useful and noble emulation in the breast of the soldier, have always been sufficient for all requirements. Punishments and rewards, based upon moral power, have a wonderful influence, and are susceptible of infinite modes of application, of which each will act strongly upon generous hearts.

Punishment, whatever it be, should never be attended by any expression of contempt, except when for an act of flagrant cowardice. Whatever degrades and tarnishes the reputation of the soldier, will likewise diminish his valor; while everything which exalts him in his own estimation will add to his good qualities. There are a thousand different ways to vary the expression of such sentiments; a skilful chief will always choose, with discernment, the means best adapted to impress the particular character of the soldier he wishes to act upon, and as are suitable to the circumstances attending the case.

In some armies severity is pushed to excess for omissions, the extraordinary reprobation of which, when looked upon with the eyes of reason, appears quite futile. Without designing any censure, I can not approve the importance which is attached to it. When only concerning some slight matter of dress, or some momentary breach of immobility under arms, too severe punishment is not reasonable; but when inflicted with moderation, and when viewed as affecting the soldier's manners, its aim is useful. Then, a spirit of order and respect for regulations will really be shown; and, as it is with education and the formation of the habits of a lifetime, they will be always adhered to. A soldier whose deportment is defective may undoubtedly fight as well as one whose conduct is beyond reproach; but, less exact in the performance of his daily duties, he will, most probably, be less submissive to the voice of his chief.

The existence of an army is a matter so astonishing and artificial, that nothing which contributes to give to its customs a habit of order and submission can, without peril, be neglected. But it is the duty of the chief to look upon matters in a proper light, and not to exaggerate their importance.

It is of the greatest moment that commanders should seek, with particular care, to inspire their soldiers with confidence; without this close bond nothing can with certainty be expected. In times of repose, and during peace, the regular power is easily respected and obeyed; but in the midst of disturbances, which in periods of danger are always sure to come, the slightest natural obstacle may become insurmountable. At such a period confidence in one's self and others—this powerful, innate voice—gives an extraordinary degree of energy, which will lead to success.

The chief must, furthermore, provide for the well-being of the soldier; he must know, on important occasions, how to partake of his sufferings and privations; he must watch over the maintenance of order and discipline; punish when necessary, and seize with eagerness every occasion to bestow rewards; but these rewards must be just, since the reliance of a soldier upon the justice of the chief is the very basis of the degree of esteem and the sentiments which his subordinates entertain for him. The instinct of the soldier is very skilful in discovering whether a chief be worthy of his regards or not. Then severity has nothing in itself which could either frighten or wound, it being the representative of legal power; and power, when the sincere interpreter of the laws, ensures to every one the efficacious protection of his rights. Even those who experience its harshness feel, in their inmost heart, that it is necessary, salutary, and to be respected.

While maintenance of order should be at all times a subject of constant attention on the part of chiefs of all grades, love for their soldiers should not be the less deeply graven into their hearts. I have already said that this meritorious class of men, so severely treated if we consider the entire aspect of their condition, can not be governed with too much affection. They are obliged to accustom themselves to every kind of privation; their lives are but so many sacrifices; the most beautiful portion of their existence is passed in the midst of toilsome labors, and in dangers which incessantly recur; and, in spite of all this, are devotedly attached to their chiefs, if those latter will only love and esteem them. Soldiers are good by nature. If their attainments do not give to them the right of being placed in the front-ranks of society, they would merit it through the sentiments by which they are animated. The habit of order imparts a greater degree of morality.   Life full of dangers develops the noblest instincts of the heart, and inspires a sentiment of devotion—a sentiment which is a heavenly gift.   Returned to his home, the soldier is nearly always the very example of that portion of society in which he is called upon to live.   I have seen him, in the midst of disorders and atrocities, sometimes the fruits of war, distinguish himself by acts of saintly piety and of evangelical charity.* Shame and disaster upon the nation which does not honor the soldier, and which does not make every effort in its power to ameliorate and sweeten his existence !

Another duty, which should never be neglected, is to keep soldiers in the greatest activity.   It should become to them second-nature.   Like nearly all men, they are disposed to be lazy; to change this disposition is, then, equivalent to rendering them a great service.   Rest and idleness diminish the strength and lessen the courage.   Health, energy, and moral valor ordinarily increase by a life hardened with fatigues and devoted to mobility.

Military exercises are the first elements of the activity I have just spoken of; but they are not such alone.   A soldier must at first acquire a most complete instruction; after he has obtained this, to occupy him with details he already knows, and from the repeated exercise of which he can derive no further benefit, will only render his profession distasteful to him.

Only great field-manœuvres, since they offer a fine spectacle, are alone constantly to his taste; but his interest may be newly awakened by exciting in his breast the desire to excel in plays of different kinds. He might likewise be employed upon important public works, and, as a recompense, the history of those regiments who have contributed to their execution might be associated with them by being named after the regiments.   Thus grand and splendid works could be economically erected, and at the same time those ideas of glory and immortal greatness would thereby be developed which can not be too much fostered in the warrior's heart.

In the course of my military life I have never permitted any occasion to escape me which presented an application of this principle, and I had certainly reason to be satisfied, as much with the immediate success as with the increased health and spirit of the troops.   But I took

---

* I could cite many such traits; I now only recall one. During the campaign of Egypt a village revolted; a military execution became necessary to set an example. The village was set on fire, and nearly all the inhabitants were executed.

A soldier who, undoubtedly, had fulfilled his task of cruelty, was struck at the sight of a child stretching its arms out toward him. He took it up, procured a goat to nourish it, carried the child for eight days, the goat following him, when he found an Arab woman who adopted it.—*Note of Author.*

care never to go beyond certain limits, and not to compromise in any manner the military spirit of the soldiers, the conservation and development of which should never cease to be the aim of all the efforts of a commander.   Egypt, Holland, and Dalmatia still show such monuments of our past greatness and former manners.   In the last-named country two hundred and forty miles of splendid roads, through the wildest countries and in the midst of the greatest natural obstacles, have left to the inhabitants honorable testimonials of ours that will never perish.   Inscriptions graven into the rocks no doubt still tell the traveller that these works were executed by such regiments, and under such colonels.   And whenever these brave soldiers, whose remembrance is so dear to me, quitted their spades to take up their weapons, with what spirit they appeared upon the field of battle!   What strength and energy they showed during the longest marches and the greatest fatigues!

As a complementary means for the formation of troops, I would place in the first rank the establishment of large camps of instruction. They alone give to troops, in times of peace, that habit and degree of instruction most suitable to them.   Military spirit is only developed in the midst of the dangers of war, and in those assemblages which may be called its image.   Camp life, with its activity and mixture of the different arms—that peculiar state of existence so little resembling civil society—is the very element of glory and success, which can only be created by an assemblage of some duration and in the midst of public prosperity.   I do not speak of momentary gatherings, which sometimes occur in different countries, and the object of which is rather to offer a magnificent spectacle than to give instruction and to develop the qualities of soldiers; but I mean such camps as existed during my younger years, and from which came the finest and best army of modern times, and which, if it ever be equalled, can never be surpassed; I allude to the army which was encamped for two years on the coast of the British channel, and which fought at Ulm and Austerlitz.

By force of this example, and convinced by my reflections, I would wish that such permanent establishments were formed in provinces but indifferently cultivated—as, for instance, the Champagne—and that durable barracks for thirty thousand men would be erected.   They should be occupied during three months, or a less time, by the same troops.   Three similar establishments would suffice to give and to preserve the military spirit of the French army, and to impart to it such a degree of instruction as would constantly keep it in a fit state for war. But we have at the present time a still vaster ground for military exercises—Algeria—which, if it costs us dearly, repays us even more richly, when its influence upon the matters of which I have just spoken is considered.

I can not finish this chapter without entering into some detailed account how regular armies were formed in Europe. It is curious to perceive how much former armies differed, in regard to their composition, from present armies. The results drawn from this comparison will, by themselves, be presented to our reflection.

After the invasion of the barbarians and the destruction of Roman power, every special kind of military organization had disappeared from Europe. During many centuries armies had no other basis than that of feudal institutions. When experience had amply demonstrated the weakness of such temporary unions of men, assembled in haste, and without any rules, which were suddenly broken up—either owing to the caprice of the lords or to their necessities—and which thus rendered any operations, dependent upon calculations, impossible, people began to think of creating powerful bodies, regularly and permanently organized.

The sovereigns, though invested with the right, had no real power over their vassals. To free themselves of this dependence, and as soon as the state of their finances permitted them to do so, they entertained the desire to keep troops themselves, and thus originated the so-called Free Companies.

But regular revenues being necessary to constantly maintain troops under arms, and, on the other hand, regular revenues being only obtained by a system of order and a certain administrative organization, the creation of armies became thus, at the same time, the cause and the means of the beginning of civilization.

Since, however, feudal rights gave to the lords the sole disposition of their population, the former were very far from favoring the establishment of troops designed to overthrow their power ; and the sovereigns, being able to dispose of their private domains alone, thus greatly restricted and reduced to voluntary enlistments, obtained the same by means of money.

The disorders existing throughout Europe, the constant wars waged, and the multitude of little sovereigns that existed, rendered the people miserable, and the profession of soldier was presented to them as a resource. The morals of the times, besides, led each one to the indulgence of the hopes of unlimited ambition.

A soldier in those days could aspire to everything ; and all his plans had but one object—personal interest. He was not moved, as the soldier of to-day, by the only thought of discharging a duty he owed to his sovereign : to defend his country, and to acquire renown, that noble reward of public opinion, so glorious in our times. Private and captain both wished for riches, and oftentimes carried their pretensions as far as a sovereignty. The Visconti, Sforza, Escales, Ecelini, and many others, had no other origin ; and before them kingdoms had already been the spoils of Norman adventurers. To facilitate the execution of their projects, sovereigns were always compelled to employ as agents such military men as had acquired renown and reputation, and who, having devoted themselves to the profession of arms from early youth, had a large acquaintance among men capable of assisting them, and who were disposed to unite their fortunes with them.

Every one of those men, in his sphere, had his clients, and regiments were raised by contract and competition.

Ferdinand II sent for Wallenstein[2] and demanded an army. The conditions were discussed and the treaty concluded. The latter called upon the officers in his confidence and asked for their regiments, associating them with himself as participants in his own benefits. The latter conferred with their captains, who undertook to find soldiers and form companies; and the army was created. In like manner, nowadays, a sovereign negotiates a loan with some rich banker, who distributes the larger portion among his correspondents, associating them with the profits he counts upon; and these again look for the money which they may need in the pockets of capitalists.

It is easily seen that a similar organization gave to the colonel who had formed a regiment the character of ownership of the same.

From this comes the name "proprietary," which they received and retained in Austria, where, although having become regiments belonging to the sovereign, as in every other country, they have still somewhat preserved their primitive character, with their peculiar constitution and privileges. The system there pursued harmonizing, as it does, these institutions with the interests and customs of the state, offers, at the same time, a fit and noble reward to generals whose lives have been illustrated by glorious services, and guarantees to the sovereign the choice of proper officers and a good *esprit de corps*.

We can now properly appreciate the vast difference presented in the composition of our armies and those of the past centuries. Our armies are created by forced recruiting. It is so in every Continental state except in England, where particular circumstances explain the maintenance of a system which exists no longer anywhere else. The armies of our time are too numerous, and voluntary recruiting would not supply their wants; besides, there is no longer so large a mass of men existing whose necessities would force them to engage in military service; public order, which now everywhere is reigning for the benefit of humanity, has greatly diminished their number. Lastly, the chances of fortune in the military career are now too restrained to actuate any valuable member of society voluntarily to engage in military service in any such manner. Other outlets are opened for the development of industry to every one possessing perseverance and intelligence, and fortune may now be acquired without the risk of danger. Forced enlistment appears then the only way of creating the necessary means for defending the state; and thus it is that a tax upon blood has become one of the public burdens everywhere.

The spirit of armies has been greatly lessened through it; but, despite appearances, it is far from having lost anything. Voluntary recruiting, with terrible discipline, may sometimes, as in England, have given good troops; but can we compare, in point of morality, an army composed of the sons of families elevated by a spirit of order and obedience to the laws, to one which perhaps has in its ranks some individuals who are animated by a love of war and of glory, but the great majority of which is composed of vagabonds, whose bad morals keep them strangers to a quiet and laborious life?

How much better is public interest secured, if those to whose hands it is entrusted look upon military service as a noble and important duty! The young man whose lot it is to be conscripted, peaceful as he is in his habits, may leave his family with regret, and with pain even; but a warlike spirit, so natural to mankind, and particularly to Frenchmen, will soon inspire him. He will henceforth entertain noble ideas; he will be exalted in his own estimation, and he becomes a faithful and devoted soldier, who finds in the good opinion which his commanders and companions have of him a reward for all his sacrifices, labors, and dangers. Such is the European soldier of to-day—for the system is alike everywhere.

It remains still to be determined which is the preferable system of the two: Whether to place in the same regiment the recruits from the same portion of the country, or to distribute them into different corps. The first one is adopted in Austria, Prussia, and Germany; the second, in France and Russia. Each has its advantages and its difficulties; but I am in favor of the first system.

To commence with its difficulties. This system gives to the soldier local and provincial ideas, which, after so many revolutions through which we have passed, would not be without danger, easily to be foreseen, should like events again take place; perhaps it likewise diminishes military spirit in times of peace, and tends to make a union of peasants rather than of soldiers: but these difficulties may be easily remedied if the number of assemblies be increased, and the duration of the time devoted to camps of instruction be prolonged.

As to the advantages, they are great and incontestable. With regard to the administration, the recruiting is more easily accomplished; the officers of the corps have the means to superintend the men absent on leave, and the transfer from peace-footing to war-footing is wonderfully simplified. In the moral aspect, to the sentiments of honor which make every soldier a part of the glory of his own regiment, and individually responsible for the same, we add, at the same time—and the effect is by no means unimportant—the obligation of defending the reputation of the province in which he is born.   It is an additional motive and a new encouragement.

Furthermore, a soldier who has distinguished himself is rewarded for his good conduct by the esteem which he enjoys in his corps; while, by the system followed in France, he is deprived of this advantage upon his retirement from service.   Returned to his home, his achievements are no longer known; he thus loses that prize held highest in his life— the enjoyment of the good fame he has won.   It would follow him, on the contrary, if, amid the scenes of his childhood, he could gather around him the companions of his youth; he would then be surrounded by the halo which his well-merited and conquered honors shed around him to the day of his death.*

NOTES.

1. **Success in organizing the Confederate Army—General Samuel Cooper, C. A.—Confederate soldiers and United States soldiers.—** The most remarkable instance, without doubt, in modern times, of the rapid and successful raising of an army, under most difficult circumstances, is to be found in that of the Confederate States.

In Europe, where military organization is so complete—where standing armies have been since the time of Charles VII of France—where everything needed for the raising and equipping of large bodies of men, in all branches, is so abundant, and where, even in the most peaceable times, a great nucleus of old soldiers may always be found—the fact of raising a large army presents nothing remarkable.

Even in times of exhaustion, as was the case with Prussia in 1813, or with France in 1814 and 1815, there was still a powerful nucleus around which every man able to bear a musket could be rallied.

The instance which may be likened to the Confederate States was the rapid formation of a Polish army in 1830, complete in its organization; but there again we find a nucleus of thirty-two thousand men, splendidly organized and disciplined, who at once embraced the cause of their country.

That which would mostly seem to resemble the case of the Confederacy appears the Army of the United States, but, upon examination, nothing could be further from the truth.

The United States had over twenty thousand regular troops.   The large dépôts of arms and ammunition, the great military and naval arsenals of construction, built up with Southern gold, were within their territories.   The navy, without the exception of one man-of-war, was theirs.   The trade with the world stood open to them.   Manufactories of every kind were at work to fill the requisitions of the army.   What they could not themselves supply, England was only too glad to furnish.

---

* The Council of War in 1828 was occupied by this question.   General d'Ambrugiac, one of the most distinguished officers of the army, the reporting member of the Committee on Infantry, had presented a mixed system which, by creating an excellent reserve, solved the question in a perfectly satisfactory manner.   Fate ordained that nearly all the labors of that council, in which military questions were profoundly and carefully debated, came to no conclusion.—*Note of Author*.

19

The Confederate States found almost nothing with which to create an army, save their heroic sons, and about a hundred thousand, mostly old-fashioned, muskets; no artillery, no ships, no manufactories.   Everything had to be created and to be developed.   There was no nucleus around which to rally, save the love of country and the few old and tried officers who have since raised the name of the South so high as to be the theme of admiration upon the lips of every tyranny-despising man.   The commerce of the world was shut to us.   What was freely granted to our enemies, to us was denied.

No people has ever been more willing to sacrifice everything for the cause of liberty than ours; but an army could not be created and organized without experienced men, and, happily, they were found.

Out of a very chaos of men, with valor certainly, but with only little notion of discipline, by a wise system of gradual tuition and organization, armies have been made, the most wonderful structures of human society.   For now three years they have gallantly checked the invaders upon a hundred bloody fields; but they still exist. A population of less than six millions of people have now had for three years an army which bore upon its lists at no time less than half a million of combatants; the fourth year is approaching, and still, to-day, as we are assured by the highest authority, an army confronts our foes which is as strong, as brave, better disciplined, and as ready to die for the country, than any we have had before.   The men may be often ragged and barefooted, but their guns are shining, their bayonets are sharp, and they have plenty of ammunition.

In Richmond there is a department, so quietly managed that a passer-by would hardly suspect its existence, but for the large number of uniformed men which may be seen to come and go.   Over this department presides General Samuel Cooper, Adjutant and Inspector-General of our armies—one of the chief agents who have achieved all this.   With a talent for organization unrivalled in the annals of war, he has discharged the tremendous task devolving upon him.   President Davis happened once to be the Secretary of War of the United States, and he went out of office with the reputation of having been the most efficient they ever had.   There he acquired, no doubt, that large experience of judging men in that army, for which he has since been remarkable.   There he likewise learned the value of General Cooper.   True to the principles of liberty upon which the former Union reposed, General Cooper could not be brought to violate them.   He, therefore, joined us, giving up high rank, friends, and home, for an uncertain future, but for a glorious cause; and from the day he was appointed the first general in our army his labors have been unremitting and severe.

To discharge his duties, justice, delicacy, firmness, intense application, and sagacity were required.   General Cooper has been successful in each; and the highest praise which can be given to any public officer he has won: namely, there is, perhaps, no officer of the army who is so universally esteemed and respected by those under him as he.   He is no longer in the prime of manhood, but still in the full vigor of mind.   In military service for now fifty years, he, nevertheless, may be seen hard at work early in the day and late at night.   The application to duty of one so advanced in years would shame thousands of men not one-half his age.

When the names of those shall be enumerated by a grateful posterity who, by their wisdom and experience, have guided our people to the goal of independence, General Cooper will not be the last, for his services have been among the first.

He has been likened to Berthier, the celebrated adjutant-general of the first Napoleon.   He may be compared to that marshal as regards his capacity for organization; in every other respect he is vastly superior.

When we consider the chief difference between the Confederate and United States' armies, nothing could better express it than the remarks of Marshal Marmont in the preceding chapter: "Can we compare, in point of morality, an army composed of the sons of families elevated by a spirit of order and obedience to the laws, to one which, perhaps, has in its ranks *some* individuals who are animated by a love of war and of glory, but the *great majority of which is composed of vagabonds, whose bad morals keep them strangers to a quiet and laborious life?*"

The character of the Federals did not stand very high long before this war had prominently brought out their most abominable traits. It is curious now to read some remarks which have been made, long ago, to illustrate it. Dr. Robert Jackson, for instance, in his "View of the Formation, Discipline, and Economy of Armies," a late inspector-general of English army hospitals, and who had ample opportunity of judging, having been attached to the English forces in America during the Revolutionary war, gives a very unimpassioned account of his impressions. He does not think them "Capable of leaving their homes for houseless liberty in the woods. The majority, particularly those whose ancestors, if not banished from Great Britain for evil deeds, had been adventurers in pursuit of fortune, had little attachment to the country unconnected with its productiveness;" hence, he says, "It is reasonable to believe that it was more in irritation from violence committed on property by arbitrary taxation, from hopes of getting rid of British debts, or from a factious spirit among themselves, than from a real desire of independent liberty," that they revolted. "They talk boastingly when danger is distant; they are not in general bold and resolute when the hand of power grasps closely." He believes them to be "easily affected by fear" and a resolute front; but "always ready to bribe themselves," they are "sufficiently shrewd" whenever this practice is attempted upon them.

"Though not daring in close combat, they were not without courage. It was a courage of circumstance, the direct combat: front to front was supported with resolution, the retrograde was precipitate when the flanks were turned, when the design of turning them was discovered, or when a front attack was threatened by the bayonet. Accustomed to circumvent, and to shoot from behind cover, they were themselves afraid of being circumvented; and impressed, perhaps, with the idea of circumvention, they moved off precipitately at the appearance of suspicious manoeuvres being practiced against them; they had not, as a soldier ought to have, a face for flank and rear."

The Federals, like the Russians, then, appear, in the opinion of Dr. Jackson, not proof against the bayonet; and, like the Austrians, have always a wholesome dread against rear attacks; two hints which can not be too much acted upon by Confederate generals.

Agreeing with Marmont, he pronounces them "Speculators after gain, rather than patient and industrious as simple laborers." He considers them deficient in two qualities that are essential to the formation of military force, "Namely, the subordination which submits patiently to such forms of moulding and discipline as renders the human race a machine, obedient to the will of a general, to whatever point it may be directed, or to whatever purpose it may be applied; or, secondly, the ardent love of country which, rising to enthusiasm, produces acts of individual heroism beyond the calculations of tacticians, and superior to the acts of mere mechanism."

An English writer very recently compares thus the Northern and Southern armies:

"The Southern army, compared with that of the North, is supposed in Europe to be physically inferior. That opinion, so far at least as it applies to the troops met with along the route from Culpeper Court-house to Richmond, is a mistake. Tall, straight, muscular, the Confederates are in general as fine a material for war as any men in the world. These Virginians, particularly, make a magnificent soldiery. One of the most marked differences between the two armies is that between the men's faces. The countenance of the rank and file, on each side, differ so much as to present a strong contrast. The stolid expression which one observes in faces at the military posts along the Baltimore and Ohio railway, compares unpleasantly with the expression of the frank, genial, intelligent countenances in the ranks of the South. The distinction is, to a very great extent, one of class, but is referable partly to differences of race. The Celtic and Teutonic casts are not so pleasant to behold as that of the Anglo-Saxon; and the Virginians, save so far only as they partake of a Huguenot mixture, are of almost purely the same stock as that of England.

" The Virginians are British in their blood and in their habits. Their sympathies have always been strongly conservative and English. In the time of Cromwell they protested against the usurpation of the Parliament, in their declaration to support the Stuarts—the ' Old Dominion.' "

It would be a profitable and useful task to elucidate the chief characteristics which distinguish the Northern and Southern soldier; but it is beyond the scope of these notes. We only trust that an abler pen will undertake it.

But we can not omit to remark the exposition of our enemy's cruelties, in the masterly Address of Congress to the People of the Confederate States. It should be scattered broadcast all over the world, and in all languages. A portion of it we can not neglect to cite here:

First, the picture of our army:

" Our army is no hireling soldiery. It comes not from paupers, criminals, or emigrants. It was originally raised by the free, unconstrained, unpurchaseable assent of the men. All vocations and classes contributed to the swelling numbers. Abandoning luxuries and comforts to which they had been accustomed, they submitted cheerfully to the scanty fare and exactive service of the camps. Their services above price, the only remuneration they have sought is the protection of their altars, firesides, and liberty. In the Norwegian wars the actors were, every one of them, named and patronymically described as the king's friend and companion. The same wonderful individuality has been seen in this war. Our soldiers are not a consolidated mass—an unthinking machine—but an army of intelligent units. To designate all who have distinguished themselves by special valor, would be to enumerate nearly all in the army. The generous rivalry between the troops from different states has prevented any special pre-eminence, and hereafter, for centuries to come, the gallant bearing and unconquerable devotion of Confederate soldiers will inspire the hearts, and encourage the hopes, and strengthen the faith, of all who labor to obtain their freedom."

And, then, the picture of the enemy's atrocities:

" Not content with rejecting all proposals for a peaceful settlement of the controversy, a cruel war of invasion was commenced, which, in its progress, has been marked by a brutality and disregard of the rule of civilized warfare, as stand out in unexampled barbarity in the history of modern wars. Accompanied by every act of cruelty and rapine, the conduct of the enemy has been destitute of that forbearance and magnanimity which civilization and Christianity have introduced to mitigate the asperities of war. The atrocities are too incredible for narration. Instead of a regular war, our resistance of the unholy efforts to crush out our natural existence is treated as a rebellion, and the settled international rules between belligerents are ignored. Instead of conducting the war as betwixt two military and political organizations, it is a war against the whole population. Houses are pillaged and burned. Churches are defaced. Towns are ransacked. Clothing of women and infants is stripped from their persons. Jewelry and mementos of the dead are stolen. Mills and implements of agriculture are destroyed. Private salt-works are broken up. The introduction of medicines is forbidden. Means of subsistence are wantonly wasted to produce beggary. Prisoners are returned with contagious diseases. The last morsel of food has been taken from families, who were not allowed to carry on a trade or branch of industry. A rigid and offensive *espionage* has been introduced to ferret out ' disloyalty.' Persons have been forced to choose between starvation of helpless children and taking the oath of allegiance to a hated government. The cartel for exchange of prisoners has been suspended, and our unfortunate soldiers subjected to the grossest indignities. The wounded at Gettysburg were deprived of their nurses, and inhumanly left to perish on the field. Helpless women have been exposed to the most cruel outrages, and to that dishonor which is infinitely worse than death. Citizens have been murdered by the Butlers, and McNeils, and Milroys, who are favorite generals of our enemies. Refined and delicate ladies have been seized, bound with cords, imprisoned, guarded by negroes, and held as hostages for the return of recaptured slaves. Unoffending non-combatants have been banished or dragged from their quiet homes to be immured in filthy jails. Preaching the Gospel has been refused except on condition of taking the oath of allegiance. Parents have been forbidden to name their children in honor of ' rebel' chiefs. Property has been confiscated. Military governors have been appointed for states, satraps for provinces, and Haynaus for cities."

2. Duke Wallenstein.—Wallenstein, Duke of Mecklenburg and Count of Waldstein, is a name conspicuously distinguished in the military operations of Europe during the early part of the seventeenth century, especially in the " Thirty Years' War." The family of Waldstein had belonged to the nobility of Bohemia for many centuries, and the hero of this memoir was born in his father's Castle of Hermanie, in September, 1583. In his youth he pursued his studies at Pavia and Bologna, where he acquired an extensive knowledge of languages, mathematics, and other sciences connected with the military art. Waldstein, anxious to signalize himself by military deeds, went to Hungary, and served in the imperial army against the Turks. After the Peace of Sitvatowk, in 1606, Waldstein returned to Bohemia, and married an aged but wealthy widow, who died in 1614, and left him large estates in Moravia. In 1617 he raised a body of two hundred dragoons, with which he assisted the Archduke Ferdinand of Austria, who was at war with the Venetians. In a short time he saw himself at the head of several thousand men; and, after the successful conclusion of the campaign, toward the end of 1617, the Emperor Mathias made him his chamberlain and colonel in his armies, and created him count. Immediately afterward he married the daughter of Count Harrach, and the emperor, on this occasion, conferred upon him the dignity of a Count of the Holy Roman Empire. The States of Moravia appointed him commander of the Moravian militia; but on his refusal to join the Bohemians against the emperor they deprived him of his command, and confiscated his estates. Waldstein was now appointed quartermaster-general of the imperial army, and in the course of the following campaign, by his timely relief of Boucquoi, who was attacked by Counts Mansfield and Thurn, near Tegu (10th of June, 1619), he saved the emperor from being made a captive in his own capital. It seems that the resources of the emperor being exhausted, Waldstein gave large sums for the support of his master. After the overthrow of King

Frederick of Bohemia the estates of his adherents were confiscated, and the reward of Waldstein was the lordship of Friedland, and other property of immense value. Waldstein was neither intoxicated by his triumph nor by his wealth. In 1621 he took the field against Betlen Gabor, the Prince of Transylvania, and forced him to sue for peace, which was granted on condition that he should give up his claim to the crown of Hungary. During the two ensuing years Waldstein was principally occupied with the management of his estates; but Betlen Gabor having again taken up arms against the emperor, Waldstein hastened to Hungary, and arrived just in time to save the imperial army, which was besieged in the camp at Goding, on the frontiers of Moravia. As a reward for this victory the emperor, toward the close of 1623, conferred upon him the title of prince, and in the following year (1624) created him Duke of Friedland and Prince of the Holy Roman Empire. On the declaration of war of the Union of Lower Saxony, headed by Christian IV, King of Denmark, which put the emperor into great embarrassment, Waldstein raised, at his own expense, twenty-eight thousand men, with whom he marched toward the Lower Elbe. The renown of his military skill, his wealth, and his liberality was so great, that men flocked to his camp from all parts of Europe, whom the iron hand of their commander kneaded into a well-united mass. The results of this campaign, so glorious for the imperialists, belong to the history of the Thirty Years' War. The campaign was begun and finished in 1626. Waldstein lost twenty thousand men by disease and fatigue; but in the beginning of 1627 he was again at the head of fifty thousand men. On the 1st of August, 1627, he was at Troppau. On the 30th he took Domitz, in Mecklenburg, after a rapid march of two hundred and fifty miles. The Danish war was finished by the Peace of Lubeck (13th of May, 1629). Waldstein's reward was the Duchies of Mecklenburg. He chose Wismar for his residence, and obtained from the emperor the title of Admiral of the Baltic and the Oceanic Sea (the German Sea). His plan was to form a navy with the assistance of the Hanseatic towns, and to prevent Gustavus Adolphus, the King of Sweden, from choosing Germany for the theatre of his ambition. No sooner was Waldstein invested with Mecklenburg than his numerous secret enemies changed their calumnies and intrigues into open accusations. Maximilian, Duke of Bavaria, and Tilly, were among the most powerful of his enemies. Maximilian at length declared to the emperor that he and all Germany would be ruined if " the *dictator imperii* " remained longer at the head of the imperial armies. Ferdinand, after long hesitation, dismissed Waldstein from his command in 1630, at the very moment when Gustavus Adolphus left the coast of Sweden for the invasion of Germany. Waldstein then retired to Bohemia, and resided alternately at Prague and at Gitschin, where he lived with such splendor as to make the emperor himself jealous. The empire was on the brink of ruin, and there was only one man who could save it. This man was Waldstein. When he at last yielded to the supplications of the emperor to resume the command, he showed that he felt all his importance. Among his other conditions he demanded that he should have sovereignty of the provinces that he might conquer, and that the emperor should give him, as reward, one of his hereditary states (Bohemia?), of which he should be the sovereign, though as a vassal of the emperor. The campaign of Waldstein against Gustavus Adolphus was unsuccessful. After losing the Battle of Lützen he punished with death many generals, colonels, and inferior officers, who had not behaved well in the engagement. He soon repaired his losses, and his arms were victorious in Saxony and Silesia. But his haughtiness became insupportable, and he openly manifested his design to make himself a powerful member of the empire. His old enemies, among whom was the Duke of Bavaria, now conspired against him. They represented him as designing to overthrow Ferdinand's power in Germany, and the emperor was the more ready to believe the accusation as it became known that France had offered to Waldstein to aid him in obtaining the crown of Bohemia. The emperor ordered him to withdraw from Bohemia and Moravia, and to take up his winter-quarters in Lower Saxony (December, 1633); but Waldstein neither would nor could obey his order, which he regarded as a violation of the conditions on which he had resumed the command. Finding, however, that the emperor was resolved to dismiss him, he prepared to resign the command. His faithful lieutenants urged him not to abandon them, and, in order to prove their invariable attachment, they signed a declaration at Pilsen, 12th of January, 1634, in which they promised to stay with Waldstein as long as he would be their commander. This is the famous declaration which has always been represented as a plot against the emperor. Piccolomini, Gallas, and several other Italian and Spanish officers availed themselves of the occasion to ruin Waldstein, and induced the emperor to sign an order by which Waldstein was deprived of his command and declared a rebel. Piccolomini and Gallas were commissioned to take Waldstein, dead or alive. Waldstein ultimately took refuge in the Castle of Eger, whence he tried in vain to negotiate with his enemies. It was through the treachery of some of his own officers, who had been bribed by the emperor, that he was destined to die. On the 25th of February, 1634, the commandant of Eger gave a splendid entertainment to Waldstein's officers, at which the duke was not present, on account of his ill-health. After dinner an armed band rushed into the room, and the friends of Waldstein fell beneath their swords. Captain Devereux, at the head of thirty Irishmen, then rushed into the apartment of Waldstein, who received his death calmly from the hands of their leader, as he stood in his night-dress, in an utterly defenceless state.

CHAPTER II.

MILITARY SPIRIT AND DIFFICULTIES OF COMMANDING.

A mystery—Danger—Instinct of preservation—Relative activity—Soldier and general have the same sentiment—Nobility of profession—Friendship—Fear—Discipline—Imitation—Rewards for true courage—Bravery—Three degrees—Which most rare—Its rewards—Reciprocal confidence—New armies not possessed with it—National guards—Appreciation of the enemy's character—Napoleon's real genius—Confidence, discipline, instruction, necessary to constitute an army—Two different kinds of soldiers—Their relative value—Why do we go to war—Command—Profession and genius—Authority—Predestination—Napoleon—Qualities requisite for the command—Will—Mind, character—The predominating element—Equation—Historical names—Decision—Counsellors *ex officio*—Painful duty of command—Responsibility—Resumé—Inferior commands—Note—General who is also sovereign—General Bonaparte has more merit than Emperor Napoleon—Parallel—Whom to prefer for command, sovereign or general.

NOTE.—Calmness of generals after having made all dispositions for the battle.

A union of one hundred thousand men in the same spot, far away from their families, property, and interests: the exhibition of their docility, obedience, mobility, and state of preservation; finally, the existence of a spirit which animates them in such a manner as to lead them to throw themselves with pleasure into the most imminent dangers, in which many of them will find death itself, at the mere signal of a single man—is assuredly one of the most extraordinary spectacles which could be presented in the society of mankind; it is a phenomenon, the cause and principle of which can only be found among the mysteries of the human heart.

It is a part of our nature to seek and love emotions; the idea of danger pleases us, although there are but few men who, at the most threatening moment, would not be disconcerted. But we feel the necessity of comparing ourselves with other men; emulation is natural to us; every one loves to believe himself and to see himself superior to his associates. Such is the motive power, by virtue of which our instinct of conservation gives way to the noblest impulses of courage.

The sphere of activity in which self-love moves depends upon the situation of the individual. Every one wishes to be seen and admired. Man, placed in a crowd of men, finds his horizon in the immediately surrounding mass; in a more elevated position, the extent of the horizon is enlarged; and he who attains the very pinnacle, looks down upon the world.

This sentiment, so honorable to mankind, inspires the most generous actions. It is the stimulus in the simple private's breast as well as in the general's. Thus, in every grade, is the profession of the soldier a noble one, since it consists of sacrifices for all, which are, beyond anything else, rewarded by public esteem and glory. To speak disdainfully of those who compose the great body of armies, is nearly akin to blasphemy; to speak of them with indifference, is not to understand the conditions of our own nature.

The sentiment I have just painted resembles and is compatible with another sentiment equally noble—namely: friendship.

Common dangers, glory, and common interests, establish the strongest and sincerest bands; and as everything is connected and united in the great mystery of society, it is precisely in a state of war and in the midst of perils—that is, when society stands most in need of such bands—that friendships are most habitually formed; or, in other words, that habit of companionship, the *esprit de corps*, to which public opinion attaches so much importance.

An exchange of rendered services, some reciprocal aid received or given, doubles, nay augments tenfold, the strength and security of each individual. Thus, a strong appreciation calls forth, develops, and exalts man's virtues in measure as circumstances render their practice more necessary, whenever self-preservation renders their exercise of importance.

But the heart of man is ever changing, and the best sentiments are combated by others springing from the same principle, but only considered in a different light.   I brave a danger in order to save my comrade, because I count upon him in a like instance; but should the immediate danger appear to me too pressing, and fear be superior to the interest by which I am attached to the individual in danger, the instinct of my future preservation vanishes in my eyes before the power of the present peril, and, fleeing from danger, I forget all the motives which should have led me to brave it. The sentiment which determines my conduct in this instance, and which is called fear, is not of rare occurrence in face of a real danger; it is even more common than could be believed, and exercises a greater influence in the case of a large body of men. It is, therefore, precisely in order to oppose this, and to strengthen contrary sentiments, that the power of discipline has been called to the assistance of authority; and as example exercises a great influence upon mankind, and since brave men, above all others, often carry away their comrades, those who act differently from the common rule can not be too much rewarded in every possible way, in order to carry their generous dispositions to a still higher degree; because, upon these very men, the fate of battles oftentimes depends.

Bravery, in the present armies of Europe, and particularly in the case of officers, may be classed thus:

Bravery which prevents an officer from dishonoring himself, and which leads him to perform his duty rigorously; it is not of any rare occurrence.

That which urges a man even beyond the requirements of his duty; it is much less common.

Finally, that which decides a man to risk his life without hesitation in order to ensure a success to which he has been charged to contribute; it is of the rarest occurrence. Thus, whenever this degree of bravery is shown, honors, riches, and great consideration should be its reward; and the opportunity of bestowing such recompenses being limited and scarce enough, there can be no fear that the expenses might become a heavy charge upon the treasury of a state.

The sentiments of which I have just spoken are not the only ones which should find a home in the breast of military men. To impart to troops the whole valor of which they are susceptible, it is necessary that confidence exist among all those composing an army. The soldier must believe in the valor of his comrade. He must be convinced that his officer, equally brave, is superior to him in experience and in instruction; he will suppose his general to be as brave as any, and possessing much more science and talent. If this be the case, then the army forms a bundle of fasces which nothing can break. In it we must seek for the first requirement of the strength of armies, and the first element of success.

But this fundamental principle, which we call confidence, is only inherent in old and tried troops, and not in new troops, who do not know each other as yet.   Hence the absurdity of a system of national guards designed to replace troops of the line.   Even supposing national guards to consist of the bravest men in the world, they will, nevertheless, be worth nothing at their first trial—since the valor and the capacity of each individual can only be appreciated by the others after some actual experience.   The first encounters, therefore, will be made without the assistance of confidence, and will most probably lead to great and irreparable disasters.

For a general the entire moral aspect of war consists in the knowledge of the movements which animate the minds of military men, and in the correctness of his judgment of the same, combined with the application he makes of knowledge thus obtained in the various chances of war.   This knowledge must embrace both the troops of the enemy he has already once met and expects to engage, as well as his own soldiers.

Here is, then, a faculty, independent of the profession proper, and it is nothing less than the inheritage of genius.   All great generals have possessed it; and never any man in the world in a higher degree than Napoleon.

Discipline, which is the auxiliary of courage, is likewise necessary as a means of order.   Its whole importance is at once felt when we reflect upon the mechanism of an army, and ask ourselves how a like multitude is able to exist, both when moving and when at rest.

To constitute an army it is not sufficient to unite men in a greater or less number; they must also be organized.   I have already explained by what mechanism obedience is secured—namely: by bringing him who commands, according to the different grades of military organization, into contact with a limited number of men, upon whom his powers are easily exercised.

As soon as this division was set at work and the organization completed, discipline commenced its operations; that is to say, the subordinates were accustomed to a passive deference toward their superiors.

After it came instruction.

Thus we see three different operations are necessary to make of a mass of assembled men an army:

1. To organize them;
2. To discipline them;
3. To instruct them.

And the complement of organization, of discipline, and of instruction, is confidence.   It is the essential element, whose absence deprives an army of the greater part of its valor.   This confidence should extend to all and each one; soldiers should have confidence among each other in all reciprocal relations; and every soldier and officer should confide in the superior chiefs, and, above all, in the supreme chief.

This precious element, which influences every result in so powerful a manner, increases proportionately in its effects with the greater intelligence of the soldiers—because this confidence, being founded upon a knowledge of men and things, is not a sentiment which has been but indifferently considered, nor a blind faith.

Soldiers without intelligence have but little mobility, and vary less than others who are more lively and greater reasoners. The command of the former is more easily exercised, and there is less risk in giving it to a general of but limited capacity; the latter soldiers, on the contrary, will have more or less valor, with the greater or less worth of the general placed at their head.

In speaking of these two kinds of soldiers, I have especial reference to the Germans and French. The Germans have often obtained successes with very mediocre chiefs; the French are worth ten times their number if they are commanded by a chief whom they love and esteem.

They will be below all comparison with a general who can neither inspire them with esteem nor with confidence. They have proven this at Hochstett, in 1704; before Turin, in 1706; and in 1813, at Vittoria. The reason of this is a simple one. War is not made for us to be killed! The object is always to conquer the enemy; and if one runs the risk of dying, it is conditional that the sacrifice of one's life to which one submits may be of some use. Whenever the moment comes that an intelligent mass of men see no probability at all of victory before them, there is no further chance of a glorious engagement; from that very time they will hesitate to compromise their lives, and seek to preserve them for a time when the sacrifice shall be attended with some useful result.

I have sought to explain the divers movements which pass in the soldiers' heart—movements which, in the opinion of the ignorant, appear to result in contradictory phenomena, and who, by considering men as so many passive machines, do not at all compehend the variations of which they are susceptible. I come, now, to the consideration of the command, and I will essay to establish its necessary qualifications.

The art of war is composed of two distinct parts: the profession proper, and the moral part, which is an appendix of genius. I have already considered war in its moral aspect, and I will but add a word or two as to the qualities which give authority to a chief over those surrounding him.

There are some persons who have an innate faculty of acting upon the minds of others, a natural power of exerting their authority, which makes obedience to them an easy matter. This authority is a particular gift, and springs from hidden causes beyond the comprehension of our mind. One who obeyed yesterday and is called upon to command to-day, handles his power, from the very moment even in which he becomes invested with the same, with as much facility as if he had always been entrusted with it. On the other hand—and the examples are frequent—another exercises authority over his equals which, though not contested, does not seem to repose upon any right, and he may not be endowed even with any superiority of mind; this faculty is one of the effects of a proper organization. The legal chief, possessing this faculty, imbues those who have to obey with salutary fear. He passes for a severe man, and the very severity which is supposed to be part of his character makes its application unnecessary. One glance, a single word, acts upon the minds of others with irresistible ascendency. These men are specially selected by Providence to command.

But as such natural and powerful influences over equals are but rarely met with, it has been necessary to prepare the structure of obedience by accustoming subordinates to honor and to respect their chiefs. This structure is composed of military grades; they determine the rights of command, and give a social position, distinct and constant, to those who are invested with it. Upon elevated grades public honors have been conferred, to act upon the mind and to speak to the imagination. Finally, nothing has been neglected to exalt these depositaries of power in public opinion, in order to secure obedience the more readily. Obedience is easy in ordinary times, when no obstacles oppose themselves to preserve regularity and order, but it becomes difficult in times of danger, suffering, and passion. When the general has the reputation of courage and ability, esteem and confidence are at once excited in the breasts of his soldiers, and his power over them is thereby augmented; should he join to this the great consideration of illustrious birth, by which his social position is a very elevated one, he will be still more exalted in the eyes of the multitude. The greater the power and reputation of him who is invested with authority, the more easily will he be obeyed.

All these requirements, being united in the person of Napoleon, greatly favored his successes. They made, if I may express myself thus, the necessary elements of the command. But what are the personal qualities which the exercise of command requires?*

---

* I have above stated the qualities most favorable for the exercise of command; and it results therefrom that, when a general is at the same time the sovereign, everything conspires to come to his assistance. As, for instance: absolute liberty in his projects, movements, and operations; accumulation of means and resources; absence of responsibility; liberty to engage in hazardous combinations, which, with great dangers, may lead also to great successes; certainty of being always

The art of war, considered as a profession, wholly rests upon combination and calculation. I have already entered into circumstantial details when speaking of strategy and tactics. But that the combinations may lead to favorable results, it is necessary that a strong will direct them; because the change of measures already resolved upon, if the motives leading to such a change are not sufficiently explained, has many difficulties, and oftentimes results in great disasters.

Two things a general must, therefore, possess: Mind and character.

---

obeyed, whatever may happen, and to be served with zeal, etc. When compared with a position so advantageous, a simple general has only the most limited powers at his disposition. Whatever these powers, he can only exercise them within certain limits. It is not sufficient for him to perform his duties in a satisfactory manner, but he must for ever be prepared, and in advance, to justify his undertakings. Lastly, difficulties may occur in obtaining the obedience which is due his grade; and rivalries, hate, and intrigues may conspire to become as powerful an auxiliary for his overthrow as the enemy he has to fight.

The two positions can not be compared; and the merit attached to a successful general is much greater than that of a sovereign. Thus, the glory which Napoleon acquired in Germany is not at all equal to that of General Bonaparte in Italy. In the first campaigns, without name, without experience in the command of armies, with the most feeble and incomplete resources, an inferior and badly-provided army, he obtained glorious successes, conquered Italy, and maintained himself there. In the other campaigns, if we leave out of consideration the series of splendid combinations they developed, the magnitude of resources of all kinds, their accumulation and abundance alone, it would appear, should have enabled a general to gain the victory without any application of genius.

The chances of success being more numerous with the military sovereign than with a general, it appears desirable that the former should command; it is, however, different in reality.

In the first place, who will be the competent judge of the sovereign's talent? and who can be the guarantee that his illusions will not inspire him with fatal confidence? Supposing, even, that he did not assume the supreme command until after numerous trials, there will be another great danger to the state, namely: should he be unsuccessful, public confidence in the very stability of his power will be shaken—a vast social disaster! Besides, public interest requires that the commanders of armies be controlled by some other power. Whatever may be the latitude conferred upon a chief, there are certain limits which he should not be permitted to exceed; and should a general be left entirely free to act, who can warrant that he would be moderate in the chances presented to him? The greatness of catastrophes is in proportion to the accumulation of means and the extent of the enterprises, and then society may be shaken to its very foundation. The faults of a general may always be repaired in a powerful country; those of the sovereign, who is exalted in public esteem and consequence, will lead to its complete ruin. Therefore the sovereign should content himself to govern and control the administration, to create resources, and to make them abundant; he should likewise bestow unlimited confidence upon him who is worthy of it, and reward with magnificence and without jealousy; but he should never assume the charge and responsibilities of the actual command of the armies.—*Note of Author.*

Mind—because, without it, no combinations can be made; one surrenders without any defence.   Character—because, without a strong and obstinate will, the execution of conceived plans will never be secured. But here the relative qualities prevail over the absolute qualities, and character should control mind.   In this state we find the element of success.   If every quality could be expressed by figures, I would prefer a general having five parts of mind and ten parts of character, to one who has fifteen parts of mind and eight parts of character.   Whenever character has the ascendency over mind, and the latter is of a certain extent, the chances are that an object determined upon will be attained. But if mind is superior to character, the judgment, projects, and the direction are being continually changed, because a man of vast intelligence considers questions at every instant in a new aspect.   If strength of will does not interpose between these continual changes, we float for ever between two parts, irresolute which one to take.   We end by choosing none whatever (which is worse still); and, instead of approaching our object, an uncertain march brings us further from it, and we are lost in a wilderness.

It would, however, be a very wrong conclusion to suppose that much mind is not requisite to achieve great things.   We nowhere meet with any generals who possessed a mediocre mind in ancient and modern times, among any of the names which have become historical—Alexander, Hannibal, Scipio, and Cæsar.   They all possessed the most distinguished faculties of intelligence.   The same we find to be the case with the great Condé, Luxembourg, the great Eugene, Frederic, and Napoleon.   But every one of these great men joined to superior qualities of mind a still higher degree of character.

This necessity of possessing a character outbalancing the mind, is felt by him who is called upon to command, at almost every instant; because, in that position, it becomes frequently of moment to take a determined part and to come to some speedy decision.   Hence that which men deprived of character dread most is to arrive at a determination; a fatal instinct leads them to postpone a resolution, oftentimes of great urgency, and which, when at last made, is of no further avail because it was delayed, and sometimes even becomes disastrous.

This remark authorizes me to proclaim this principle: A general is justified in receiving counsel when he feels the want thereof; but to be habitually led by counsels, unless he be compelled to do so by the supreme chief, can never be attended with success.

The necessity, therefore, of coming to a resolution is the most painful duty of a commander.   In that moment responsibility rises before him with its imposing cortege—all the duties with which he is charged, and all that is dear to him in the inmost depths of his heart, are vividly before his mind.   The responsibility toward those dependent upon him, the discharge of the responsibilities assumed toward the public welfare, toward himself, and his conscience; in its complete aspect, a responsibility so immense and terrible in proportion as the general is penetrated with the consciousness of the duties devolving upon him. There is but one way to support this weight; he must have enough strength and resolution to place himself above all consequences—sure to find, in his conscience and his intentions, a generous approbation of all his actions, after having devoted to them all the capacity and the intelligence of which he is capable. But there are very few men able to rise to such a height. This necessity of coming to. a decision is both of the greatest importance and of the greatest difficulty to him who commands. But whenever the part has been chosen so as to admit of no further modification, and the cannon begin to thunder, when the battle has begun, and every one has received the indication of the part allotted to him in the opened drama, then a commander-in-chief may rest tranquil. He has again found the security and repose of mind of which he was deprived before the battle opened.

We have seen, then, that a general unites all the qualities promising success, when he possesses the mind to see, judge, and combine, and the character to execute; when, furthermore, he has a complete knowledge of men, their leading passions, and their secret motives, of which .war calls forth so many; when, besides, danger, far from paralyzing his faculties, causes them to augment and to show forth with renewed energy; when, finally, he loves his soldiers and is beloved by them, and when he unceasingly thinks of their preservation, their interests, and well-being as a father should of his children. I have only said that success is then promising, and not secured; because war has such varied chances, and is subjected to so many hazards, that nothing is certain before actual success has been completed.

In treating of the necessary qualities for the exercise of command, I have intended to speak of the chief command. Any other command, however extended it be, can not, at the moment of its becoming subordinate, be at all compared to a chief command, however limited the latter may be, considering the number of troops—because there the same difficulty, which I have sought to make comprehended, does no longer interpose, namely: the necessity of resolution. I have commanded under Napoleon both armies of different strength and *corps d'armée*. Only ten thousand men, left to the sole control of their chief, require much more solicitude, and leads to much greater embarrassments, than the command of fifty thousand men belonging to an army of two hundred thousand men. In the last case they are moved, march, and fight according to given orders and a fixed object—comparatively easy matters; and whenever the engagement or the march is terminated and the camp is established, the general can repose as the lowest soldier does, and has nothing to do but to await orders. At that very time, on the contrary, is the commander-in-chief harassed the most by troubles and providential measures of all kinds.

NOTE.

Calmness of Generals after having made all dispositions for the Battle.—This calmness, so well described by Marmont, appears to have been one of the chief characteristics of all great generals.

Frederic the Great, after his dispositions had been made, frequently threw himself upon a bed of straw to take some rest, or, seated before a camp-fire, would quietly muse and fall asleep.

Napoleon never rested until even the slightest matters had been carried out to his satisfaction, and during the night preceding the battle would himself receive all the reports; when this was done, he was soon enjoying a few hours' rest in the midst of his faithful guards, until Berthier would wake him to mount on horseback on the morning of the battle.

Just before the Siege of Rodrigo, when the proximity of the allies to Marmont's army placed them in considerable danger by reason of the non-arrival of their flank divisions, a Spanish general was astonished to find Wellington, the English commander, lying on the ground in front of his troops, serenely and imperturbably awaiting the issue of the peril. "Well, general," said the Spaniard, "you are here with two weak divisions, and you seem to be quite at your ease; it is enough to put one in a fever." "I have done the best," the duke replied, "that could be done, according to my own judgment, and hence it is that I don't disturb myself either about the enemy in my front, or about what they may say in England."—*Edinburgh Review*, July, 1859.

In General A. S. Johnston and Lieutenant-General Jackson, of our army, we remark the same characteristic feature.

CHAPTER III.

PICTURE OF A GENERAL WHO ANSWERS TO ALL THE REQUIREMENTS OF THE COMMAND.

His bravery—His calmness—His *sang-froid*—His dash—He must have an established reputation—If not, he must eagerly seize the opportunity to establish it —His physical qualities—Privileges of youth when called to command—He knows the value of time—His activity is without limit—He must be everywhere —His severity—His sentiment of order—Kindness united to severity—Merited punishments—He must respect the uniform of soldier—His gravity—Does not exclude affability—Facility of access to him—He must attend to everything at once—His magnificence—In what it ought to consist—His topographical knowledge—Campaign of Marengo—Fort of Bard—How to deceive the enemy—How to discover the enemy's intentions—The general's entire independence necessary—A general not to be absolutely dependent upon the government.

A review of the life and character of the late General Albert Sidney Johnston, Confederate Army.

I will here, in a few words, group together the qualities and the bearing which should characterize a general called to a chief command.

He is a brave man, and recognized as such by the whole army; his courage can not be one moment called in question, or become the object of the slightest doubt. His bearing is characterized by calmness and *sang-froid*, without, however, excluding that dash and impetuosity so well calculated to inspire and carry with it those who witness the same. If his reputation in this respect be not sufficiently established, he should seek and seize the opportunity of giving to it an incontestable basis; otherwise he will never be able to exercise upon his generals, officers, and soldiers that power, commanding respect and esteem, which is so indispensable to success.

This reputation once achieved, he must avoid being prodigal of his life, without, however, being too much concerned about its preservation.

His mind is, as has heretofore been established, subordinate to his character.

His bodily strength is proof against the greatest fatigues, and considerations of health never prevent him from inspecting personally the most important matters; since even the best made reports, and the accounts of the cleverest persons, will never have the same precision acquired by a personal investigation.

If nature has endowed him with superior faculties, it is desirable that he be called to the chief command at an early age. By it, his successes will be sooner secured. He will then possess that wonderful energy and confidence in himself which doubles his strength. An object of sympathy and good-wishes to every young soldier of his army, he will at the same time never be wanting in that great deference due to age, and he will be personally endowed with sufficient experience. There are matters, a just knowledge and judgment of which is only acquired by time and by experience, and which can never be divined. But, on the other hand, a too prolonged habit of obedience diminishes, rather than develops, the qualities requisite for command.

It is especially necessary that he should have seen war when very young, and shortly after entering upon his career; if this be not the case, he will only slowly and with difficulty acquire that tact and instinct which an early acquaintance with war creates, and by which its difficulties are singularly simplified.

He will continually be impressed by the fact that a surprise can only be the consequence of culpable negligence, and that a surprised general is dishonored. By guarding against derelictions of duty, he will not only shelter himself, but all of his subordinates, against any reproach.

Knowing the value of time, the only treasure which can never be again recovered, he will dispense with writing much, by leaving it to those in whose sphere of duty it is to transmit his orders; he will only reserve to himself their revision and approval. A good general has

20*

never written much in the field; the head and not his hand should be busy; he employs his time more usefully by giving verbal instructions, and to turn it to account by preserving the freedom of his judgment to inspect whether his intentions have been carried out in the proper spirit, and in the meditation of new combinations.

His activity must be without limits: by his often unexpected presence he will keep every one in fear of being caught when wanting in the discharge of his duties, and in this way he will keep alive the zeal of all.

Every one of his decisions will be ruled by impartial justice; he will be severe in the maintenance of order and discipline; thus warranting to the soldiers the enjoyment of their rights, and the greatest degree of welfare compatible with their situation.

If severity is one of his duties, there is another one, discharged with much more pleasure, and which is not less important. I allude to the rewards due to glorious actions and good conduct. He must not be prodigal nor parsimonious in their distribution; he must make it his own business to obtain them; and he should place even more importance in their acquisition than he would in his own personal recompense. Let him, however, reflect well that he award them to those only who have a right to their claim; because a reward justly made is an encouragement to the heart of the generous, while an unmerited recompense destroys all emulation, and is, moreover, productive of intrigues. The instinct of the men, and their innate love of justice, will quickly see through the spirit which influenced the manner of the distribution of rewards.

If the general is true to these principles, if he satisfies the conditions just enumerated, he will be the object of respect, esteem, and affection on the part of his troops. The necessity of order is so vividly felt by military men, that they always love a general whose severity is the warrant of that order; they confidently trust him, because firmness and equity characterize all his decisions.

Kindness without severity leads to nothing; it is mistaken, both by impression and in reality, for feebleness—which makes a chief the plaything of surrounding influences. But kindness, accompanied by severity, resting upon principles, makes the general the idol of his soldiers. Rigor, however great, is based upon laws, and restricted to certain limits, which should prevent it from becoming injurious. A man easily resigns himself to a merited punishment, but injury is irritating. The greater the calmness with which a punishment is inflicted, the greater will be its effect. If a chief is violent, he can not complain if his subordinates murmur. A general should also treat every one who wears a soldier's uniform with consideration. There is something so noble in their profession, the sacrifice of life is so sublime, that those who, by reason of their calling, are always ready to offer it, have a right to command consideration, even if they have merited severe punishment.

A general should be habitually grave in his manners whenever he comes into contact with his subordinates; still, these manners do not forbid a sort of familiarity and dignified gaiety, as they will inspire affection and esteem. A feeling of fraternity is naturally excited among military men by common dangers, privations, and fatigues; nor is this manifestation in any way incompatible with the regulations of military subordination and the maintenance of discipline. The less a general appears to be conscious of his superiority, the more is the soldier aware of the same. A general should be accessible to everybody.

Despatches which he receives should be at once opened, forgetful of any personal inconvenience. Should he even be roused twenty times in one night, amid the fatigues of a campaign, to receive reports of little consequence, he should never prohibit his being called up again. News, in war, may be of such importance, and a delay of but two hours may prove so fatal, that the safety of an army sometimes depends upon it.

A general's mode of living should be as magnificent as his fortune may permit. His first luxury will be the keeping of a great number of horses; he should have a number of them sufficient for any possible emergency. The second object of his magnificence should be a house, adapted for the exercise of the greatest hospitality. An officer whose duties carry him to head-quarters during war, should never leave it without having experienced proofs of his general's hospitality. It is not only an act laudable in itself, but staff-officers, or officers finding themselves at a great distance from their corps, would, if the general did not take care of them, be reduced to great stress for their subsistence, and might even be compelled to suffer. To this consideration of humanity may be joined another, which even affects the good of the service. The officer, charged with some mission, hastens his arrival, knowing in advance that he will be hospitably received. He accelerates his march because he is kindly disposed toward his chief, and for his own interest; and time is so important a consideration during war, that we should economize it as much as possible.

A general should neglect nothing to obtain, beforehand, a minute knowledge of the country into which he proposes to carry his operations. He will procure the best statistics; he will know in what the resources of all kinds consist, if he study the topography of his field of operations but carefully. The least negligence in the pursuit of this study may lead to the gravest consequences. We can not too much reflect upon all matters belonging to the characteristics of a country, and to be able to turn every resource to account. By procuring at any price the very best maps, and by looking at them continually, even sometimes in a vague manner, he will be sure to acquire happy ideas, which may become of immense value to his operations.

The insufficiency of information of this nature nearly occasioned the miscarriage of the immortal campaign of Marengo in 1800, at its very outset, and led to many difficulties. The First Consul did not know of the existence of the Fort of Bard, and its means of resistance; it would have been an easy matter for us to seize it, had we attached to the first bodies of troops some pieces of sufficiently heavy calibre. The fact that the Little Saint-Bernard, which debouches, like the Great Saint-Bernard, into the valley of Aosta, was also practicable for artillery, had been ignored; the passage of the mountains would have been much more prompt, and would not have presented so many obstacles, the conquering of which may justly be considered as one of the most remarkable operations of our epoch.

Every one of his projects demands the profoundest secresy; a general should communicate them to those only who are called upon to execute them, and not until the moment when this knowledge becomes necessary. How many well-conceived enterprises have miscarried because they were known to the enemy! On the other hand, nothing is more favorable to success than to permit a belief to gain ground, contrary to what we wish to execute; by deceiving those who surround him, a general will most surely deceive the enemy.

But it is of quite as much importance for him to inquire into the projects of the enemy as it is to hide his own; a general will neglect nothing in this respect. Without reposing a blind faith in his spies, he will well pay and entertain them. It is especially useful to procure intelligence from those employed near the enemy's head-quarters.

To succeed in this, the first one of his cares will be to know the detail of the organization of the different corps composing the army of the enemy, and the names of the commanding generals. With this aid, and that of light troops, well commanded, who, constantly engaging the enemy, make prisoners, he has important documents at his disposal relative to the enemy's movements. The capture of a single soldier of such and such a regiment announces the presence of a certain division which belongs to a certain corps, and from it a general perceives the spirit and the object of the manœuvres and operations of his adversary. It will be scarcely believed how much candor, simplicity, and truth are contained in the answers of a prisoner, without his suspecting the range of the questions he is asked, and without his believing in the least that he harms the cause he has served with zeal, and which he is very far from wishing to compromise.

Finally, the general who holds his glory dear must keep himself aloof from any absolute dependence, in any of his operations; it will always be fatal. An enlightened government does not assume the pretension to direct everything; its rôle is limited to the indication of the object, after having determined the nature of the means and their quota. To the general alone it belongs to determine the system to be pursued and the combinations to be followed, since he is placed in the face and the midst of the surrounding difficulties. Rather than submit to any too direct action of the government, the general should abandon a command which he is not permitted to exercise in its whole plenitude; it is either necessary that the government renounce the entertaining of different views from the general, and still accord him its confidence—or that the general be deprived of the command, if it be believed that he is following a bad system. The government should only act upon a general who has its confidence through the influence of counsels which do not bear the character of imperative orders; it will, above all, take care not to place near him any counsellor *ex officio:* since there is nothing more absurd than such a system; and, as I have already said, its results can not but be always fatal. A general may very properly provoke discussions, and consult enlightened persons; he may even accept counsels if he thinks it useful; but he should never be compelled to ask for and submit to them. A general at the head of an army has but two rôles to play—to obey or to command. Let the government give the command of its armies to those it believes to be worthiest, and let it accord to them, at the same time, an unreserved confidence; if not, let the generals be displaced.

A REVIEW OF THE LIFE AND CHARACTER OF THE LATE GENERAL ALBERT SIDNEY JOHNSTON, C. S. A.

I.

Two foreign officers, in the service of the Confederate States, were ordered to report for duty to General Albert Sidney Johnston in the month of October, 1861. When leaving his head-quarters at Bowling Green, in the State of Kentucky, having then seen and spoken with him for the first time, they simultaneously exclaimed, when outside of the enclosure of the unpretending quarters: "He is the very *beau ideal* of a general!"

To one of these officers, who now feebly attempts to pay this humble tribute to the memory of the departed hero, this, his first impulsive exclamation, has become the basis of the greatest veneration of which he is capable.

In the prime and vigor of genius, and upon the outset of a brilliant career, at the period of manhood when the maturity of powers has been reached, such as are attained through a course of trials and of hard service falling to the lot of the American officer of the frontier; eminently fitted for the most responsible position that man can be called upon to fill—the command of thousand of human beings—making him one of the pillars upon which the fortunes of a country reposed, and the hopes and liberties of a nation rested—General Johnston, too soon, to our clouded perception, met the death which it is a true soldier's highest honor to covet—the death upon the field of battle, when the legions which he had led were marching to victory.

Others, whose names are as dear to Southern soldiers as that of Albert Sidney Johnston, have, indeed, risen higher and higher from those cohorts which form the Army of the Confederacy, since his fall, and some have shared his fate; but there is no one whose name calls forth greater sorrow in the national bosom, or is destined to shine more illustriously, than that of the chief who fell upon the bloody ground of Shiloh.

The full character and glory of such a man can neither be perceived nor established amid the agitation of a struggle for life and liberty such as that in which we are engaged. New reputations arise every-day; a few will be lasting, but the majority (and some, it may be, which at this very time cast a shadow over the truly great) will sink into obscurity before the scrutinizing researches of serene and unimpassioned history. Asperity, jealousy, and ambition—more developed, perhaps, among military men than in any other class of mankind—will then have run their course. Now, every rising name is more or less subject to such baleful influences; and some persons (and they are the most ungenerous of all) even continue their aspersions after the victim is no more; and can we say, in truth, that the memory of Johnston, even after his form is cold and he is sleeping his last sleep, has been unassailed or justly dealt with?

We, who have witnessed the prominent men of our struggle moving upon the theatre of action, and who have been inspired through them, can only furnish the materials from which, long after we have ceased to breathe, the future historian will erect the monuments of their glory, or pronounce the verdict of their shame, and consign them to tombs of utter forgetfulness.

General Johnston was born in the old Pioneer State of Kentucky—a commonwealth which has given to us some of the most illustrious men. Appointed from Louisiana, he entered the United States Military Academy of West Point. The severe discipline and exactions of that formerly celebrated training-school of Southern soldiers (now, no doubt, suffering, alike with every Northern national institution, from the influences of a corrupt government) brought out his military traits. Unlike hundreds of cadets who, after a short probation, had to succumb and to make room for more worthy competitors, he, at the expiration of the usual period of rigid application, graduated the eighth among a class of forty-one; and when we now look at the slender list remaining, we entertain no doubt that he will be for ever the leader of his class.

Upon his leaving West Point General Scott tendered to him the position of an aide-de-camp on his staff. But General Johnston, with a true soldier's aspirations, rather sought the hardships of the camp and the dangers of warfare in preference to the easy life and continuous round of pleasures of the metropolis of North America, and left for the field as a lieutenant in the Second infantry.

From this time commences a series of arduous duties on the frontiers of the United States—operations which form a connecting link between peace and war—services which may justly be compared to the celebrated achievements of the French army in the sandy plains of Algeria, and to the hardships of the Russian army among the rocky mountains of the Caucasus. The chieftains of both these nations, many years after, met upon the shores of the Crimea, and there they practically demonstrated, when confronting each other with highly disciplined armies, how far they had profited from warfare against savage and uncivilized nations. So with us has the continuous warfare against an unfortunate but barbarous race resulted in lasting benefits to our officers, which have been amply shown in our struggle.

Continued service of this kind, however, without the inducements of distinction greater than is offered against bands of savage guerillas, had the natural effect of making the majority of officers of the late United States army neglectful of completing the structure of military science and acquirements other than mere duties of routine, the foundation of which had been laid in the military school; nor, indeed, did they have, in many cases, the opportunity of prosecuting their studies amid the ever-recurring changes from post to post, often thousands of miles from the belt of civilization. This neglect, so common among them, must have been seriously felt when their view had to embrace many thousands of men, and when their mettle was being tested by "grand operations" quite different from the contracted nature of partisan warfare. But such neglect can not be imputed to General Johnston. Indeed, his reputation for skill and learning was at that early day spreading among his companions-in-arms—a reputation gloriously vindicated by his future career.

Nor was his experience in the field, in the old service, of slender proportions. We see him mentioned with distinction in the so-called Black Hawk war, in the wide Western territories, as the adjutant of the Sixth United States infantry. In that war we find General Johnston, General Taylor, and President Davis all serving together. There, most probably, was begun that deep and lasting friendship, which, outliving all the attempts of slander and of envy, still lingers over the tomb of the noble soldier, and which throws unfading honor over him who did not withdraw the light of his trust and affection from the darkened path of a friend, upon whom were heaped the reproaches of nearly an entire nation.   As a singular coincidence we may mention that, in that same war, figured Abraham Lincoln, the valorous captain of a company of militia, distinguished for cowardice and utter inefficiency in the position he had assumed; and whose military fame, then so auspiciously begun, has since spread in so remarkable a manner that we have no further need to revert to it.

Fired by a cause which strongly appeals to every generous bosom, General Johnston, during the memorable struggle of Texas for independence, united his fortunes with those of the infant republic.   The people of Texas were not slow to discern and to employ the experience of one quite willing and ready to serve in the capacity of a private soldier for the vindication of the sacred principle of self-government; and from the position of adjutant-general of the Texas forces he rose to that of senior brigadier-general of the army, and subsequently was Secretary of War until the year 1840; thus giving four years of undivided service to a state which now points, with proud distinction, to him as the first of her honored sons.

Thus, the gallant sons of three great states—Kentucky, Texas, and Louisiana—are united with his history, and are called upon to vindicate and avenge his death.   In Kentucky stands the home of his childhood, Texas guards the spot where he had gathered around him a devoted family, and in Louisiana repose his ashes.   Alas! not even these have been left untouched by the sacrilegious hands of our enemies.   Though the fortunes of war have temporarily separated these states from the main portion of the Confederacy, their sons will still find, in the life and example of Albert Sidney Johnston, a new and lasting incentive to pursue steadfastly the glorious path they have heretofore trodden.

Upon the breaking out of the war with Mexico and the United States, he served as colonel of a Texas regiment until it was disbanded, and afterward on the staff of Major-General W. O. Butler, as inspector-general, at the storming of Monterey, in whose report of that action he was most favorably mentioned.   During the campaign he was greatly trusted by General Taylor—who, indeed, paid to the young soldier the high compliment of sending for him before the Battle of Palo Alto for consultation.   There Zachary Taylor, A. S. Johnston, and Jefferson Davis were a noble trio among the many rising men, but no soldier was more revered and trusted than Albert Sidney Johnston. He was then regarded as the great military man of the United States army.   Many reputations went down in that war; and the great glories gained by some while it lasted have since faded, and are eclipsed by the deeds of men who were then but in subordinate rank, such as

Beauregard, Joseph E. Johnston, Bragg, Lee, Davis, and Albert Sidney Johnston.

There is no task more pleasing, and yet more sad, than to turn over the leaves of official documents—now almost forgotten in the greater grandeur of the present—which illustrate past deeds of arms, and to meet the names of those who, then giving promise of great actions to come, have since been inscribed by a grateful nation, with letters of gold, upon the tablet of fame. While we mourn over the departed heroes, we, with proud and swelling hearts, bid God-speed to those who have hitherto been spared. And there we read their own modest accounts of actions which have made them famous—strongly in contrast, when we turn the pages, with those written in the spirit of self-laudation and evident consciousness of their own glories, by men whose achievements have sunk into oblivion, because there was no element of greatness to be found in them; and on other pages we encounter the records of gratified commanders, expressed in terms of praise, which now sound almost prophetic.

There we see nearly all the names which our history has since made great. There are the documents—results of laborious duties, constant dangers, severe privations, and hard studies—which have become the foundation of the edifice of martial honor, and to which the historian will ever recur with fond satisfaction. And that which strikes the reader most is the unpretending, unembellished, and modest character of these papers—the stamp of true greatness. When we contrast them with documents of a like nature of renowned commanders of other nations, the same feature may there be found; content to receive what will voluntarily be given, they lavishly bestow all credit upon those who were led by their genius alone. A true soldier, though stern and unyielding in the discharge of his duties, is not, therefore, the less modest, and has no grain of that overbearing manner which strongly marks those characters who find all the satisfaction they desire in the contemplation of their own greatness.

Every one who in the least reflects upon the character of General Johnston, must be penetrated with the conviction that this noble trait was one of his most admirable qualities. The higher he rose upon the ladder of military advancement, the more honors and friends his station gathered around him, the more is it apparent. Unmoved by the increasing responsibilities of his various positions, and the applause of friends, or the words of interested flatterers—the Scylla upon which many a fair reputation is wrecked—his modesty was the same.

Intimately linked to it was a degree of forbearance and charity, rarely attained by mortal man, and without its equal in the history of modern times. The letter to President Davis, in explanation of his abandonment of Kentucky—which, though written in private confidence to an unwavering friend, has since become the property of the nation—magnanimously uttered in the very midst of condemnation heaped upon him, will not fail to carry sorrow to our hearts for ever, and is a solemn warning to the nation, as well as to individuals, not to pass judgment with clouded vision.

In the expedition against the Mormons, for the command of which he was singled out from many distinguished United States' officers, some other of his prominent traits appeared: moderation and firmness. To them the honorable success of that expedition must be mainly ascribed. Reviewing the condition of affairs in a despatch to General Scott, commanding the United States army, he says, January 20, 1858:

"My information respecting their conduct since is, that their troops are organized to resist the establishment of a territorial government by the United States; and, in furtherance of that object, they have erected works of defence in the mountain-passes and near Salt Lake City. Knowing how repugnant it would be to the policy or interest of the government to do any act that would force these people into unpleasant relations with the Federal government, I would, in conformity with the views also of the commanding general, on all proper occasions have manifested in my intercourse with them a spirit of conciliation; but I do not believe that such consideration for them would be properly appreciated now, or rather would be wrongly interpreted; and, in view of the treasonable temper and feeling now pervading the leaders and the greater portion of the Mormons, I think that neither the honor nor the dignity of the government will allow of the slightest concession being made to them.

"They should be made to submit to the constitutional and legal demands of the government unconditionally. An adjustment of existing difficulties on any other basis would be nugatory.

"Their threat to oppose the march of the troops in the spring will not have the slightest influence in delaying it; and if they desire to join issue, I believe it is for the interest of the government that they should have the opportunity."

General Scott, who appears to have been chafed by the refusal of General Johnston to serve on his staff, did not forgive him for many years. Only after the general had been appointed colonel of the Second United States cavalry, and his splendid administration of the Department of Texas had forced General Scott to testify the regard which no man can withhold from real worth, the relations of the two soldiers became more cordial; and during the period of the expedition to Utah General Scott appears to have been a warm friend of General Johnston, and he repeatedly gave evidence of the favorable opinion he entertained of him. In a despatch directing him to take command, he says:

"No doubt is entertained that your conduct will fully meet the moral and professional responsibilities of your trust, and justify the high confidence already reposed in you by the government."

And upon some other occasion the adjutant-general of General Scott, Irvin McDowell, of Bull Run memory, writes to him:

"The general directs me to add that he has equal confidence in your judgment, discretion, zeal, and gallantry; and in the delicate and responsible duty with which you are charged he desires to leave you as little trammelled as possible."

But such was General Scott's desire to earn some further laurels, that he felt ill at ease to permit one of his able lieutenants alone to reap the harvest, and he gravely announced to General Johnston, through his aide-de-camp, Lieutenant-Colonel George W. Lay:

"The general-in-chief himself will set sail for the Pacific, in the steamer of the 5th proximo" (February 5, 1858), "clothed with full powers for an effective diversion or co-operation in your favor from that quarter."

The Mormons probably never knew in what serious danger their rear at one time was; and it must be considered lucky for them, General Scott, and the final success of the expedition, that this "effective diversion" was soon after relinquished.

Before that little army, which was to penetrate through the boundless plains of the West, to enforce obedience to civil authority, lay a distance larger and fraught with greater obstacles than was presented to the French in their memorable onward march to Russia. When compared with that marvellous host, it was indeed but a handful of men which thus was determined to brave the difficulties of a march through regions which, for the most part, were bereft of any comforts whatever, and which might have become, through the interposition of the powers of nature and the failure of supplies, as disastrous to them as the return of the French from Russia had been in 1812. Everything sustaining man and beast had to be transported over a distance of more than a thousand miles. For two years the greatest burden to which any general is subjected—the provision of bread, and many other anxious cares—were sustained by General Johnston in a manner which at once stamped him a leader entitled to the highest honors. All reports concur in pronouncing that little army to have been a model of discipline, efficiency, and confidence in themselves and their leader. Those wild and silent plains will probably never again see such a body of men; and when we compare what the character of the Army of the United States was then, and what it has become since, the contemplation excites a shudder, and forces the conviction upon us that a final dissolution is indubitable, and can not, in the natural course of events, called forth through the utter disregard of all the laws of society, be very distant. The condition of the Army of Utah, under difficulties and embarrassments such as are only presented by the existence in a wilderness, showed the administrator acquainted with even the most minute details of the service; and it gave promise of greater things to come.

Speaking of this gallant little army, and the United States army in general, Mr. Floyd, the Secretary of War, in his annual report of 1858, to President Buchanan, says:

"These regiments have accomplished within the year a march, averaging for each the extraordinary distance of twelve hundred and thirty-four miles. These marches, in the main, have been made through the uninhabited solitudes and sterile deserts which stretch away between the settlements of the Atlantic and Pacific coasts, upon routes which afforded nothing to facilitate the advance, except only the herbage which the beasts of burden might pluck by the wayside.

"Every item of supply, from a horseshoe-nail to the largest piece of ordnance, has been carried from the depots along the whole line of those tedious marches, to be ready at the exact moment when necessity might call for them. The country traversed could yield nothing. * * No disaster has befallen the army, * * and the privations, hardships, toils, and dangers to which it has been continually subjected, have been borne without a murmur."

And, in another place:

"The conduct of both officers and men attached to the Army of Utah has been worthy of all praise. The commander, Brevet Brigadier-General A. S. Johnston, who joined his command at a time of great trial and embarrassment, with a calm and lofty bearing, with a true and manly sympathy for all around him, infused into his command a spirit of serenity and contentment which amounted to cheerfulness, amidst uncommon hardships and privations which were unabated throughout the tedious and inclement season of winter. The destruction of our trains by the Mormons, the disasters which necessarily flowed from it, drove General Johnston to the necessity of sending a detachment of men to New Mexico for supplies essential to preserve the whole command from the greatest extremity, and to enable him to prosecute his march with all practicable despatch."

Again, one year after, Secretary Floyd said, in his annual report:

"I can not commend in terms too high the wise prudence and officer-like conduct of the general commanding the army in Utah. The discipline of his command is admirable, and its efficiency is unsurpassed. Much has been done through the army under command of General Johnston toward improving the roads in Utah, and to give to the public fuller knowledge of the condition of the territory."

When, finally, through his moderation, wisdom, and firmness, the difficulty was nearly settled, and the restraints of military power could with safety be relaxed, he issued a proclamation to the Mormons, in which he assured them that "No person whatever will be in anywise interfered with or molested in his person or rights, or in the peaceful pursuit of his avocations; and, should protection be needed, that they will find the army (always faithful to the obligations of duty) as ready now to assist and protect them as it was to oppose them while it was believed they were resisting the laws of their government."

Such were the noble sentiments and forbearance of a Southern soldier toward a sect of fanatics who had done all in their power to harm him, and who acknowledged no obligations either to God or man, but what they themselves had decreed. Though he wielded the terrible scourge, which at his bidding would have destroyed and carried in its train a thousand horrors, he was firm against temptations of martial fame as long as there remained a chance to perform the more glorious part of pacificator. For the interests of humanity he thus achieved everything; for himself he claimed nothing, save the honor of having scrupulously discharged his duty.

In all of his numerous correspondence he but a single time, in his official despatches to army head-quarters, alluded to his personal feelings or affairs, or asked any indulgence. When his task was completed he wrote the following to General Scott, showing thereby another shining quality—his unceasing devotion to public duty:

"On the arrival of General Harney or Colonel Sumner," he said, "I desire to be ordered to join my regiment. If that can not be granted, I request that the general will grant me a furlough for four months, with leave to apply for an extension. I have had no relaxation from duty, not for a day, for more than nine years."

II.

We come now, with hesitation, to review the latter portion of his life. With hesitation, because many things are now obscure which, in the interest of the country, can not be cleared up, although strong and convincing testimony, now available, could be brought forward. Conforming to the spirit of our remarks in this volume, we must leave, and we gladly do so, the task of revelation to those who, from their position and knowledge, may be most competent, when our independence shall have been established, and harm can no longer result by untimely criticism; nay, when Truth and History imperatively demand that nothing be withheld. We do so with hesitation, because it is felt what restraints this necessity imposes upon the desire to do full justice. And gladly would the memory of Albert Sidney Johnston have been left to slumber in the silent admiration of a nation to that glorious day when ringing bells (if, indeed, they have not all disappeared for the casting of can-

21*

non) shall proclaim the advent of peace, did we not strongly feel that it has hitherto been somewhat neglected. There is no danger that it could ever pass away; but it behooves us to bring the traits and character of our lamented leaders ever prominently forward for the emulation of our soldiers as much as the welfare of the country, and our honor requires that we judge with charity those who are living.

When the separation took place which inaugurated, on the part of the North, a relentless war, the Southern officers of the former United States army, with but few exceptions (who, whatever their achievements have been since, in a wicked attempt of subjugation, will be consigned to eternal shame) hastened to tender their swords to their native states. Among these was General Johnston, then commanding the Department of the Pacific. The anxiety is well remembered with which his arrival was anticipated, who, from the distant shores of the Western ocean, was known to be on his way to cast his lot with his native South. With the knowledge that the Government of the United States, which, through the medium of General Scott, had already offered him the chief command of an army, to be in rank second only to the Lieutenant-General of the United States—which offer was answered by his resignation— would most probably lay difficulties in his way, should he choose the easy and comfortable route by steamer from San Francisco, he at once resolved to undertake, with but few companions, the tedious journey, full of hardships, known as the overland route, and arrived safely within the territory of the Confederate States in the latter part of the summer of 1861, entering New Orleans on the last day of August. Those who, at the time of his arrival at Richmond (September 5), were in that city, will not have forgotten the deep impression which his matchless figure and noble demeanor made upon the people. He did not pass the streets without being the object of general notice and national pride—an interest so unusually awakened in republican countries. He was at once assigned to the Department of the West—an appointment received with satisfaction throughout the country.

The enemy had just been disastrously checked in his first attempt upon Richmond in July, 1861, through the native prowess of Southern volunteers—who, upon that field so well auguring for Southern arms, had begun to record the long story of their glories, bravery, and fortitude under reverses, since so splendidly illustrated. The anxiety and uncertainty felt by the people at the opening of a war, the result of which could not be foreseen by mortal man, before yet the sullenly opposing armies had met in the shock of battle, had given way to consciousness of strength; but, alas! in the train of that unexpectedly complete victory followed the exaggerated contempt for the bravery and endurance of our adversary, and a dangerous degree of confidence, penetrating alike the people and the army, and to which the reverses soon following can plainly be traced. There, at Manassas, the battle had been fought against troops from New England and the North-eastern states, descendants of men who already, in the times of Washington, had been notorious for inefficiency and cowardice in the field. Upon a different arena were shortly to be met the hardy sons of the West, by nature and education alike differing from their effeminate companions-in-arms of the East.

At that early period already the importance of the Western states and territories, intersected by the affluents of the Missouri and Mississippi rivers, loomed up in the distance. The power in men and internal resources of the enemy there threatened us most where the great valley of the Mississippi spreads, in teeming beauty, the most magnificent portion of the Southern states. By means of the streams, each vieing in grandeur with the most renowned of Europe, which descend the western slopes of the Alleghany mountains, and which, rolling through the most fertile states of the former Union, unite their waters with those of the Father of Waters, the enemy was enabled to penetrate into the very centre of our Western states. The check at Manassas had directed the full attention of the enemy to these vital points; and while we were glorifying ourselves he was vigorously preparing for the great onward movement which, temporarily checked at Shiloh, has since been fraught with never-ceasing disasters to the Confederacy, and still calls for a victorious general to drive back the hordes of the enemy across the Ohio river, and redeem what we have lost. The general who shall accomplish it will be the saviour of his country; for upon successes in that quarter alone may we build the hopes of a speedy termination of the war.

Unfortunately, the wavering course of Kentucky, which has since been bitterly repented by that unhappy state, now controlled by a despotism worse than that of the Czar, prevented us from at once seizing the mouths of two of these great affluents, the Tennessee and Cumberland rivers, both upon Kentucky's soil, in the north-western corner of the state, and as one glance upon the map will show, the key and the most remarkable strategical position of Central North America. Could these have been secured, through a spontaneous and hearty co-operation of the people of Kentucky with their Southern brothers, and the line of the Ohio been ours, this war would have been checked long ago. As it was, our scrupulous regard for the rights of a sister state was miserably rewarded, and would have well-nigh worked our destruction but for the incomparable fortitude of the South. We had then no means of arresting or impeding the naval armaments upon the Ohio, in view of the southern borders, almost within hearing of the workmen's hammers, who were to forge the chains for the people of the Confederate States and of Kentucky; and thus we were compelled to see the formidable fleet of gunboats rapidly completing which was soon to penetrate to the very heart of Tennessee, and to the confines of Alabama and Mississippi.

When our eyes were being opened, and Kentucky's neutrality became a myth, Major-General Polk, with great promptitude and the *coup d'œil* of a soldier, on the 4th of September seized and occupied Columbus, in Kentucky, on the Mississippi, to protect that river. This remarkable position was at once made impregnable against an attack by water, and its rear was secured by a camp of observation at Feliciana, thirty miles east of Columbus, and which held the enemy at Paducah in check—thus establishing the left of a line of defence. Nearly three hundred miles to the east, amid the wild mountains of Eastern Kentucky, and forming the extreme right of the line, Brigadier-General Zollikoffer had taken a position at Mill Springs, where the mountains of White Oak creek rest upon the Cumberland river, to protect the remarkable defile called Cumberland Gap, where three states meeting, the mountains effectually bar the advance to the invader save through a gorge easily defended. Upon the centre of the line which unites these two points the railroad from Louisville to Nashville passes through a country-town, Bowling Green, nestled between hills, and in the narrow valley of Barren river. This town had been occupied by Brigadier-General Buckner with a force of four thousand Kentuckians, a portion of the State Guard, the nucleus of a body of refugees, who thus early raising the standard against Northern oppression, have since made glorious the name of the Pioneer State in Confederate annals. To protect the Cumberland and Tennessee rivers, the State of Tennessee had constructed two forts. One, Fort Henry, was on the right bank of the Tennessee—the other, Fort Donelson, near Dover, on the Cumberland; both as high up as the boundary of the state permitted, hastily and unsatisfactorily constructed, and, owing to the scarcity of heavy ordnance, but indifferently armed.

This was the situation of affairs when General Johnston, on the 28th of October, 1861, assumed command of the Department of the West, with head-quarters at Bowling Green. Weeks before this event happened the people of the Confederacy generally were impressed with the idea that powerful forces had already been collected at every point of the line for a triumphant march upon the City of Louisville, the capital of Kentucky, and that nothing was wanting but that General Johnston should give the order for the deliverance of the state from Yankee thraldom.

Instead of this visionary force he found but a small body of troops, and an advance upon Louisville with much less than twelve thousand men was out of the question, especially when the enemy, roused by his presence, was already concentrating the main body of his forces in front of Bowling Green. Nothing remained, then, to be done but to make his defensive line as strong as circumstances and the means at his command permitted.

The means at his command were very small, and the difficulties of organization very great. Nevertheless, by the end of November, in addition to the garrisons established upon intermediate points of the immense line, he was enabled to concentrate some thirteen thousand men at Bowling Green. The exaggerated estimates of his strength, while checking the enemy with salutary effect, of course in the same ratio enhanced the confidence of our people; yet this could not be avoided. But urgent representations were made to all the governors immediately interested in the maintenance of the line of Kentucky, and nothing was from them concealed. The aid given was very feeblè. The State of Mississippi alone nobly responded, and sent to General Johnston a respectable force, with part of which Major-General Polk was enabled to relieve a division of well-organized and disciplined troops under Colonel (late Major-General) Bowen, and send them to Bowling Green. Brigadier-General Hardee had brought a further accession, and Brigadier-General Floyd joined with his brigade from Western Virginia.

Considerable accession to the force of General Johnston would no doubt have resulted, had the people of these different states responded to the demands of the Provisional Government, and enlisted for the war. The scarcity of arms was so great that the government did not feel authorized to arm twelve-months' men, when the war-troops enlisting could barely be supplied. But the people indulged in the most disastrous delusions, and could not be brought to turn from the contemplation of the glories of Manassas.

In these endeavors, steadily pursued, General Johnston did not neglect the demands of the hour. Everything that could be done was done promptly and vigorously. Out of a mass of undisciplined volunteers were moulded steady soldiers. The departments of the army were administered with rigid economy. The fortifications were strengthened; Bowling Green strongly defended by a cordon of detached forts; new works were erected at Clarksville, on the Cumberland, while in the rear the Town of Nashville was commenced to be fortified, should the difficulty of subsistence or other causes make a position behind the line of the Cumberland more desirable. An important railroad, easily assailable, and the only direct line of communication with Major-General Polk, was maintained.

The result of all this was unshaken confidence on the part of the troops in their commander. But what endeared him most to his soldiers was the great justice which was the basis of all his decisions, the promptness with which wrongs were rectified, and the facility of access to the chief commander, as well as the genuine cordiality and dignity with which every one was met by him. Heavy labors on forts in mid-winter were endured without a murmur, since every soldier knew that General Johnstou would never hesitate to expose himself whenever necessary. His head-quarters were a model of order, simplicity, and prompt despatch of business. His decisions to personal applications were immediate and final. His bearing was that of a knight of the olden times. The writer will never forget the shouts which greeted the general whenever the troops passed in review.

With the most vigorous exertions and appeals General Johnston, upon the beginning of the year 1862, found himself at the head of some twenty-three to twenty-four thousand troops, while the enemy was confronting him with a force consisting of one hundred and seven regiments, numbering at least sixty-five thousand men.

The demonstrations of the enemy had begun on the 7th of November, 1861, twelve days after General Johnston assumed command, against the forts of Columbus. There Major-General Polk, with greatly inferior forces, defeated Major-General Grant on the west bank of the Mississippi, by crossing his troops from the left bank—a victory which, while giving us increased confidence, so disconcerted the enemy that he suspended any further operations against that point. Another column, the vanguard of Buell's forces, next appeared in front of the centre, at Woodsonville, a point some twenty miles north of Bowling Green, on the 17th of December.

They met with our grand-guards, and were checked by the gallant attitude of our troops under Breckinridge and Hindman. Thus baffled upon two points of the line, a concentrated movement was made against our extreme right, under Major-General Crittenden, on the 19th of January, 1862, at Mill Springs. Major-General Crittenden, with quick resolution, under difficulties which deserved a better result, did not await the junction of the enemy's two columns in his front, and marched against one—when, after a gallant struggle, he was compelled to desist in his attack, and, under great hardships, crossed the Cumberland during the night.

The enemy had only been awaiting the completion of the fleet of gunboats to make demonstrations by water. Long before Fort Henry fell, in view of the disappointments to which General Johnston had been subjected, he was fully aware that his line, unless it was strongly reinforced, could not be held; and in the month of January, 1862, when one day looking with Colonel Bowen upon a map, showing the course of the Tennessee river, these memorable and prophetic words fell from his lips, when pointing out a spot marked "Shiloh Church": *"Here the great battle of the South-west will be fought."*

Toward the latter part of January General Beauregard arrived at head-quarters. He was astonished that General Johnston, with so small a force at his command, could have so long held so large a line.

In a conference he fully coincided with his plan of future operations, namely: the withdrawal from Kentucky, and the necessity of deciding the fate of Nashville and of Tennessee at Fort Donelson.

After Major-General Crittenden's defeat, events rapidly followed each other. Fort Henry, garrisoned by twenty-one hundred men, fell. The commanding officer, Brigadier-General Lloyd Tilghman (late Major-General), after having, with a few cannoniers, worked his guns to the last—thus giving to the entire infantry, under the junior commander, Colonel Heiman, a brave foreigner, the chance of escape to Fort Donelson, nobly preferring, with his stout little band, to share a long and painful captivity—surrendered when the last gun could no longer be fired. With its fall the rear of General Johnston's line was at the power of the enemy.

Major-General Grant now rapidly, on the 13th of February, had completed the investment of Fort Donelson, and commenced the attack. Of the troops at Bowling Green General Johnston had detached twelve thousand men, under Brigadier-Generals Floyd, Pillow, and Buckner to sustain the garrison. After three days of severe and obstinate defence, Brigadier-Generals Floyd and Pillow having withdrawn on the night of the 15th, with a small portion of the troops, Brigadier-General Buckner surrendered himself and the entire remaining garrison. Thus the left of the line of defence was severed from the centre.

Immediately after the fall of Fort Henry, preliminary orders had been given for the evacuation of Bowling Green and for the march of its garrison, eleven thousand, upon Nashville. The magazines, heavy armament, and the subsistence stores were quietly removed before even the troops knew that the town was to be given up. When Major-General Hardee, the immediate commander of the Army of Central Kentuky, left his head-quarters, the shells of the baffled enemy were dropping in the midst of his escort, and the last train had barely left when the advance forces of the enemy entered, fired the town, and indiscriminately let loose the horrors of war upon the inhabitants of the once quiet and peaceful country-town.

When the advance brigade of the troops marching upon Nashville, under Colonel Bowen, was within half a day's march of the city, the spirit of the troops was raised to the highest pitch of enthusiasm upon hearing the news from an aid of the commanding general, Colonel William Preston, that the brave garrison of Fort Donelson was still holding out. But when the troops entered the capital of Tennessee, on Sunday, the 16th of February, the dejection of the citizens assembled upon the public square at once told the story of the fall of the fort, and that it had already been resolved, as a matter of military necessity, to abandon the indefensible position of the city, and leave it to the mercy of the invaders.

Nashville is situated upon the left of the Cumberland, upon a slight plateau, gradually rising from the river bank. Had this plateau been unassailable from the heights which encircle it, a defence might have been made; but the attempt was useless with a force of only eleven thousand men, and the prospect of seeing fifty thousand men debouch upon the city from every road to the west, north, and east, not considering the auxiliary of the enemy's gunboats, the arrival of which was every moment to be expected.

The demoralization existing among the Tennessee troops, aided by the depression which had seized the citizens, would, moreover, have defeated any idea of defence. Before their eyes was the prospect of at once abandoning their state and all they held dear. Unschooled in the trials of war, as our troops are now, this was to them appalling. But few untried troops will remain steady under such influences, and it required all the energy of the more brave to control and keep their men in ranks. Here was apparent the unsatisfactory nature of volunteer troops, led by men elected by themselves, and who, no matter how estimable in common life, were not proof against the storms of adversity. Their bad influence over their men was felt to an extent which would have been disastrous to the whole army but for the steadiness and cheerfulness shown by the general commanding, and the vigorous exertions of those selected by him.

At this dark hour in the general's life, when from every quarter the voices of ignorant assailants were heard—when a portion of his troops, stimulated by those whose sacred duty it was to check and command them, openly denounced him—and when a hasty press did all to undermine the reputation of a man whom it should have sustained in the difficult task before him—General Johnston's character rose above all the din and clamor, and shone forth with immortal lustre. No complaint, no accusation ever escaped him; there was no weariness, no wavering, or indecision under his heavy burden. In those cold, stormy days of mid-winter, so well recollected by those who endured their rigor, when men and nature appeared to conspire against him, his mind was ever active and his vision clear; unceasingly working for the good of his country, striving to unite discordant elements, and to encourage, when all was dark to even the bravest, with an unfaltering faith in God, and a reliance that He would vindicate His servant, he conceived and determined upon that brilliant movement, by a rapid march, closely followed by the converging columns of the enemy on his flank and rear, to advance upon the Tennessee, cross the river, and behind its line to concentrate his forces with those of General Beauregard, at or near Corinth, in Northern Mississippi—a manœuvre which will associate his name with those great commanders who, in the midst of gloom, gloriously issued from seemingly insurmountable difficulties.

One of General Johnston's great qualities was the knowledge of the value of men, and the faculty of assigning to them the position most suitable to their talents—a quality, it must be confessed, of the utmost importance to a commander. This was illustrated in the selection of the officers of his staff. Unlike some Confederate generals we might mention, he sought to associate with himself men who were prominent in their profession and noted for their military attainments, and not those whose only recommendation was the personal claims they might have upon him. General Johnston's staff was composed of men who would have been an honor to any staff of any country, and such as no general need hesitate to confer with in moments of danger and in delicate situations.

When he had determined to effect a junction with General Beauregard, two of the most trusted officers of his staff, as he himself has recorded, decidedly opposed the attempt, deeming it too hazardous and impracticable. But such was General Johnston's firmness that nothing could influence him when once he had deliberately decided upon his course. We shall see how, in a similar case, he rejected advice coming from even more weighty sources.

Before him lay a distance of over two hundred and thirty miles, to be traversed in mid-winter, by troops unaccustomed to the hardships of prolonged marches and to privations, and who had hitherto, for the most part, lived in comparative ease during their career as soldiers.

At Murfreesboro', Tennessee, a halt was made, and the command was reorganized and augmented by Major-General Crittenden's division and the fugitives from Donelson. Confidence in themselves and their leader was rapidly re-established in the forces; on the 22d of February—that is, six days after the fall of Donelson—the reorganization of the army was completed and the orders issued, and the march, after all the munitions of war had been removed, was resumed through Shelbyville and Fayetteville upon Decatur, on the Tennessee, where the troops crossed in safety, and were rapidly established in cantonments along the line of the railroad to Corinth; and finally, toward the end of March, the army of General Johnston, 20,000 strong, united with General Beauregard at Corinth, which brought his force to nearly 50,000 men.

The main points of the Western campaign of 1861–62 have been here rapidly sketched for no other purpose than to fully and forcibly illustrate the disadvantages and disheartening circumstances in which General Johnston found himself from the very day when he arrived at Bowling Green, and to vindicate his memory, with reference to his conduct under these circumstances, from any doubts remaining.

Even while he was yet making his way across the Western plains, a train of events had occurred which ever afterward disastrously operated against him. We do not know which to admire most—his masterly check of the enemy's columns, while they were yet seeking the weak points of his line of over three hundred miles, and the splendid disposition he made with his handful of men by which it was covered, ere yet Forts Henry and Donelson had given way—the fall of one exposing his rear, and that of the other severing his left from the centre—or the fortitude and the resources of military genius he displayed in the midst of reverses such as alone would have been sufficient to sink any general not of the highest order, even if these reverses had not been attended by premature and criminal judgment against him on the part of his traducers. None but a general of the first rank could have maintained an army of volunteers, restored its confidence, and with it resumed the offensive, under like circumstances. History furnishes many examples where the *morale* of armies, composed even of regular troops, was entirely destroyed by causes much less potent than those which operated against General Johnston; not even Frederic II, after the great disaster of Hochkirch, displayed greater genius; but only few instances can be cited where an army was managed with more fortitude and success than that which followed the leadership of the lamented hero.

When the chief officers of his staff advised against his march to join the forces under General Beauregard, he, with the resources of a master of strategy, who with one glance embraces the great points of the campaign, and a genius unexampled in our war, had already planned the great change from the defensive to the offensive, and the glorious battle of which that march was but the prelude.

It was not, then, mere obstinacy, or any undue regard of unfavorable chances, which caused him to insist upon a movement which was to be but a link of a great strategical combination; but a rigid execution of a plan which he felt could alone save his country and his own fame; and with invincible resolution he steadily pursued the programme which, foreshadowed by his remarkable, well-authenticated words to the late gallant Major-General Bowen—"*Here the great battle of the South-west will be fought*"—ripened into life, and became a monument of history—alas! at the expense of a life too precious for the nation to be sacrificed for even so magnificent a battle.

Nor was his sole aim the mere possession of the battle-ground of Shiloh. His vision reached further. On the morning of the eventful 6th of April, when he was informed that the enemy had permitted himself to be surprised, he said to a staff-officer, with now significant import, "*To-night we will water our horses in the Tennessee.*"

Clear and full, like a map in his mind, as he had ordered, that battle was developed as a game of chess. With overwhelming forces he overthrew everything before him. At a critical moment, when the enemy offered an obstinate resistance, and when the possession of the contested point became the turning-point of the battle, the last reserves had been brought into action, and it became a sacred duty, at any cost, to restore the battle, his knightly form was seen leading his troops to the combat. "Fix bayonets!" rang his clear voice; onward they charged, and the field was won.

During the latter portion of March the troops occupied the chief points of the Mobile and Ohio, and Memphis and Charleston railroads, which unite at Corinth, Mississippi, where head-quarters were established; the right was at Iuka, Mississippi, eight miles from the Tennessee river, under command of Major-General Crittenden; the centre at Corinth, some twenty-two miles from the river; and the left rested upon the Memphis road, still further from the stream. This line protected the Gulf states from any further advance. Still, various attempts were made by the enemy to turn our right, by attacking the batteries of Eastport, which, however, were promptly checked by the forces of Major-General Crittenden and Brigadier-General Breckinridge.

The enemy, in the meantime, had concentrated a heavy force, under Major-General Grant, on the left bank of the Tennessee, near Pittsburg Landing, opposite our centre, threatening Corinth, with the intention of awaiting the arrival of Major-General Buell, who, by forced marches, was hastening to effect a junction with Grant. In perfect security against the formidable opponent they deemed to have entirely discomfited, they reposed upon the beautiful banks of the river, leisurely awaiting the command of the senior general hastening to their support.

A change—"one of the most delicate operation of war," as Napoleon has said—was here determined upon by General Johnston: the transition from the defensive to the offensive, against an enemy flushed with success.

Now, for the first time, he had an army with which he was confident he could teach a lesson to the enemy. With the junction of his force to the disciplined corps of Pensacola, under Major-General Bragg, and the troops of General Beauregard and Major-General Polk, full confidence animated every regiment of the army, and it burned for the opportunity to hurl back the invaders. As soon as the preparations and the labors of organization could be completed, he had resolved to march upon the enemy, to surprise and defeat him near the river, and, with a victorious army, to meet Buell. With the zealous co-operation of his generals, the different columns were reported ready on the 1st of April.

General Beauregard, to whom the immediate command of the troops had been offered, declined on account of his ill-health; but Major-General Bragg consented to take upon himself, in addition to the command of his *corps d'armée*, the arduous duties of chief of the general staff. The army was divided into four corps, commanded by Major-Generals Polk, Bragg, Hardee, and Brigadier-General Breckinridge, respectively—the corps of the latter acting as a reserve.

The three first-named corps marched from Corinth, the last from Burnsville, a point between the centre and right of the line, upon Farmington. The corps of reserves, having the longest march to perform, upon roads made impassable by drenching rains which had overtaken the troops in bivouac, found almost insuperable difficulties to arrive in time at the common rendezvous at Monterey, and, in fact, could not reach there before twelve hours after the appointed time. The artillery of Brigadier-General Breckinridge, fast in the mud, was only relieved after great difficulties by large detachments sent to the rear from the regiments composing the corps. The perplexities were so great that Brigadier-General Breckinridge reported his situation to the general. "Let a new road be cut" was, according to Major Hayden, the laconic reply the messenger received.

The attack was to have been made on the morning of Saturday, the 5th of April, and the troops were ordered to march from Monterey—a few homesteads, surrounded by woods, and some eleven miles from the river—at three o'clock in the morning, But a heavy rain falling during the night upon worn-out troops, retarded the preparations for the march of the army until about seven o'clock. Then, when in serried ranks and upon many lines, overhead gloomy clouds charged with rain, in the morning mist, at the head of every regiment, the general's *last battle-order* was read.

"Soldiers of the Army of the Mississippi," said he, "I have put you in motion to offer battle to the invaders of your country. With resolution and disciplined valor, becoming men fighting as you are for all that is worth living or dying for, you can but march to decisive victory over the agrarian mercenaries who have been sent to despoil you of your liberties, your property, and your honor.

"Remember the precious stake that is involved. in this contest; remember the dependence of your mothers, your wives, your sisters, and your children is upon the result.

"Remember the fair, broad, abounding land, the happy homes, and the ties that would be dissolved and desolated by your defeat.

"The eyes and hopes of eight millions of people rest upon you. You are expected to show yourselves worthy of your race and your lineage; worthy of the women of the South, whose noble devotion in this war has never been exceeded at any time.

"With such incentives to brave deeds, and in the trust that God is with you, your generals will lead you confidently to the combat, fully assured of ultimate and glorious success."

When the reading was concluded there rose from every line such successive shouts of determination and patriotic devotion, as gave unmistakable evidence of victory.

But, owing to the difficult march over rough roads, and the immense host assembled in one spot, delays occurred which retarded the arrival of the rear column at Mickey's house, six miles from the river, until near four in the afternoon. General Johnston had advanced on a personal reconnoissance within two miles of Shiloh Church, and would have begun the attack that evening but for the greatly fatigued condition of the troops. In such proximity to the enemy it was greatly to be feared that he would become aware of and prepare for the impending danger.

In the evening of Saturday a council of war was held. Several commanders called the attention of the commanding general to the long delay of thirty-six hours, which should have given ample time to the enemy to receive the shock; and one of the general officers there assembled strongly urged a retreat. But General Johnston decided upon and ordered the attack for the coming morning.

He followed thus the maxim of Napoleon, that "When once the offensive has been assumed, it must be sustained to the last extremity. However skilful the manœuvres, a retreat will always weaken the *morale* of an army; because, in losing the chances of success, these last are transferred to the enemy. Besides, retreats cost always more men and material than the most bloody engagements, with this difference: that in a battle the enemy's loss is nearly equal to your own, whereas in a retreat the loss is on your side only."

The wisdom of General Johnston's decision was apparent the following morning, when Major-General Hardee, with his corps, surprised and overthrew the advance forces of the enemy. From that moment he never doubted a complete victory, and the speedy recovery of all the territory he had lost—a result destined to remain unfulfilled by the interposition of the hand of death.

We do not here design to record the character, events, failures, and consequences of the Battle of Shiloh, but only wish to confine ourselves to those facts immediately bearing upon the illustrious name heading this paper. For this purpose we are sure we could offer no more acceptable and interesting account than that which flowed from the pen of Colonel, now Major-General, William Preston, the intimate friend of General Johnston, given in a letter to the general's son, Colonel W. Preston Johnston, aide-de-camp to President Davis, dated "Corinth, April 18, 1862."

"The country from Corinth to Pittsburg," says Colonel Preston, "passes over low and swampy lands, poor and uncultivated, to Monterey, eleven miles from the former place. The road then passes northward to a farm-house called Mickey's for about four miles, and a number of country roads, through hilly and wooded uplands, some seven miles to the Tennessee river. Owl creek flows nearly east into the Tennessee near Pittsburg, and Lick creek in a general parallel direction about five miles distant to the south. The ridge dividing the small

22*

branches and tributaries of these creeks lies from Mickey's north-eastward to the river, and country roads traversing hills becoming bolder and more difficult as you approach the river, pass by Shiloh, a little country chapel three miles from Pittsburg. Occasional fields and cabins intervene, but the clearings are not numerous or extensive. The enemy were encamped near Shiloh, before Pittsburg, on the verge of some woodlands, half a mile from the river, and near the fields. * * *

"The morning of the 6th of April was calm, bright, and beautiful. We were in the saddle before the dawn was clear, and a fire between skirmishers opened in the front on the line of Hardee's advance. Between dawn and sunrise sharp volleys were heard, and the general, with his staff, rode to the verge of the wood near a field where Hindman's brigade was suffering under a heavy fire. Some of the men were breaking ranks, and there were many dead and wounded. The general, in person, rallied the stragglers, and I rode forward, where I found General Hindman animating and leading on his men. He informed me that he desired support, and, having reported it to the general, he requested me to order General Bragg to advance. General Bragg, when found by me, stated that the order had been given ten minutes before.

"General Hindman pushed on in the direction of the advanced camp of the enemy, occupied by the 13th and 18th Wisconsin regiments, and other troops, from which there was a heavy fire of musketry and artillery.

"General Johnston then passed to the left at a point in front of the camps, near two cabins, subsequently used as a hospital. A field of an hundred acres, fringed with forest, extended to the north-east. Through this General Cleburne's brigade moved in beautiful order, and with loud and inspiring cheers, in the direction of the centre of the advanced camp. Heavy firing was heard as they neared it.

"General Johnston then went to the camp assailed, which was carried between seven and eight o'clock. The enemy were evidently surprised. The breakfasts were on the mess-tables, the baggage unpacked, the knapsacks, arms, stores, colors, and ammunition abandoned. I took one stand of colors from the colonel's tent, which was sent by me, next morning, through Colonel Gilmer to General Beauregard.

"General Hardee reported his men still advancing at this camp about nine o'clock, and conferred with General Johnston, who was reconnoitring a second line of camps near the river, where the enemy were posted in force. They then commenced shelling the first camp, apparently attracted by the presence of the staff and escort; the distance being, I should think, six or eight hundred yards, and shells from the gunboats, of large size, were thrown. General Johnston received a report and rough draft at this time from Captain Lockett, stating that the enemy were strongly posted on the left in front of our right. Heavy musketry firing and cannonades indicated that Bragg and Hardee were successfully advancing on our left.

"General Johnston rode down the hill to escape the shells, and his escort back toward the woods. This was about half-past nine. After pondering a little while he determined to bring forward Breckinridge's reserve, and, feeling his way to the river, to turn the enemy's left. The brigade of Chalmers was moved to our extreme right; Bowen's next, eight hundred yards in rear of Chalmers', and Statham's eight hundred yards in rear of Bowen's, in an echelon of brigades. Statham's brigade, under the immediate command of Breckinridge, then assailed the camp near the river, when they were vigorously met, and a fierce struggle ensued.

"General Johnston then deployed Bowen's brigade, and advanced to the support of Breckinridge. Batteries were brought forward, and Chalmers' extended on the right to the river. The enemy's left flank was completely turned. A few minutes afterward he was struck by a ball, and passed on. His horse was wounded in two places, and a minie ball severed the artery of his leg; but still riding on, concealing his wound, he fell at length from exhaustion. Governor Harris was near him. I found him a few minutes after he was shot, and asked him to speak to me. I could find no wound on his body. He breathed for a little while, but did not speak, or recognize me, and expired without a pang—his countenance bearing the same noble serenity in death that it had done in life.

"After his death we bore his body back to the camp, concealing his death, and reporting to General Beauregard, with whom we remained until the close of the day; and the remains were afterwards conveyed to New Orleans and deposited in the Cemetery of St. Louis."

*　　*　　*　　*　　*　　*　　*　　*

Thus fell a soldier who united, in a remarkable degree, the attributes of a general as he should be; and who offers, in the completeness of his character, the most worthy subject of study and emulation to the Confederate army.

Thus fell a citizen in the defence of his country, unswerving in his devotion and patriotism. Just, pure, and good, his name will become a household-word in every family; and as years roll onward, and much that now is will be obscured or forgotten in the mists of the past, his eminent virtues will grow more and more in brilliancy.

Thus died a martyr in the defence of what nations hold most dear, and which it is most wicked to undermine and attack—the sacred right of self-government—the independence of one's country—the security and sanctity of our homes, and all that man loves and cherishes—without which life is but a burden, and death a heavenly favor.

CHAPTER IV.

REPUTATION OF GENERALS.

Marshal Brune—Three great epochs of his life—Switzerland, Holland, Italy—Despite exceptions, merit should be judged by success—Classification of generals—The first class—In antiquity and modern times—The second class—Is more numerous—The third class—Generals composing it are rare—The fourth and last class.

Note.—Marshal Brune.

I will terminate this work by some reflections upon the reputation of generals, and the reasons which should establish the same.

Generals sometimes attach their names to successes to which they are strangers; these are successes obtained either despite their bad dispositions, or in consequence of received and properly-followed counsels.

I have known several which come under this category—among which the most notable is Marshal Brune,* who, judged by this standard, was of great mediocrity. Nevertheless, his name is associated with three glorious souvenirs: With the successes of the French army against the Swiss, in 1798; against the English and Russians, in 1799; and against the Austrians in Italy, in 1800.

In Switzerland, the superiority of our forces and the divisions existing in that country, necessarily decided the question in his favor. In Holland he was not at the engagement of Berghen; the Battle of Bewervich was brought about by accident, and suddenly fought without any plan of conduct or any aim. The follies and the stupidity of the Duke of York alone led to any definite results. In 1800, in Italy, after brilliant successes, to which the general-in-chief was nearly a stranger, we were in a condition, had we but had another man at our head, to destroy the hostile army entirely.

On the other hand, we could cite examples where the efforts of men of great talent were wholly unattended by fortune.

But these different examples do not prevent us from judging by results, and this is the justest manner of appreciating the value and the merit of generals.

To assume a different basis, and to support one's judgment by the opinion alone which one has formed of a general's mind and talents, would be to enter into an inextricable labyrinth, and would often lead to error—since every one would then but look through the prism of prejudice, friendship, and passions. If we are sometimes mistaken by judg-

---

* Despite his private qualities and his deplorable end, we can not but point out Marshal Brune as one of the most singular and striking instances of the caprices of fortune.—*Note of Author.*

ing upon the basis of facts, we will err much more frequently by solely reposing our judgment upon a personal knowledge of the individuals. Fortune may once or twice overwhelm a man with its favors who is not worthy of them, and it may betray the highest combinations of genius and humiliate a noble character; but when the strife is prolonged and events are multiplied, then a man, complete in all the requirements of a general, will infallibly succeed; and if continuous reverses succeed each other we may boldly conclude that, despite a superior mind and qualities which have dazzled us, one defect of harmony in his qualities will destroy the charm.

I may class generals in four categories.

In the first I place generals who have gained every battle they fought. The very first place in public opinion incontestably belongs to them. But their number is so small that we can barely find their names. In antiquity I see but Alexander and Cæsar. The Grecian generals whose names have become illustrious, as Miltiades and Epaminondas, owe their celebrity to one or two actions.

In modern times I see only Gustavus Adolphus, Turenne, Condé, Luxembourg, and Napoleon until 1812; because I place, with reason, among the number of reverses for which a general is accountable, the destruction of armies, the cause of which was a want of care and an exceeding improvidence.

In the second category I place generals who, having often gained battles, sometimes lost them after they had disputed them obstinately. They are among those whose greatest number is inscribed in the temple of memory. Perhaps there are some among them worthy to figure among those of the preceding class; because, between two equally meritorious generals, while the victory must be decided in favor of one of them, it will be but a dearly-bought one, and its results will be limited.

In the third category will be those generals who, habitually unlucky in war, and sustaining frequent reverses, have never been destroyed or discouraged, but have always maintained an imposing and respect-inspiring front toward the enemy.

These generals are rarely met with, since they must have a great ascendency over those who surround them. Such have been in antiquity Sertorius and Mithridates, and in modern times the celebrated Wallenstein, and William III, King of England.

Finally, in the fourth category, we naturally find those who lose their armies without fighting, and without making the enemy win his victory through a vigorous resistance. Their names can be easily recalled, for each country and every epoch has furnished instances of them.

NOTES.

Marshal Brune.—Guillaume Marie Anne Brune, born on the 13th of May, 1763. This officer entered the French army in 1791, and was at once appointed adjutant-major. He continued attached to the adjutant-general's department until 1797, when he was promoted to the rank of division-general. In this distinguished position he successively commanded the Armies of Italy, Holland, the West, and the Reserve. In December, 1801, he was appointed a state councillor, and in September, 1802, was sent on an embassy to Turkey. On the 19th of May, 1804, General Brune received a marshal's bâton. In the following year he commanded the coast army, and in December, 1806, he was appointed Governor-General of the Hanseatic towns. During the Empire he passed through the degrees of the Legion of Honor; and as he declared his adhesion to the Bourbon cause after the Battle of Toulouse, he was, in 1814, honored with the cross of a Chevalier de St. Louis, created a peer of France, and governor of the eighth military division in 1815. He did not long survive the fortunes of his great master, Napoleon, or his own new honors. In August, 1815, he was brutally murdered by an infuriated mob at Avignon, instigated by some malicious leaders, who accused him of having been an active participator in the massacres of 1792, when, in fact, he was at the time serving on the frontiers.

CONCLUSION.

From all that precedes, I believe we are able to draw the following conclusions:

1. The fundamental principle of the organization of an army is found in the united spirit which influences an assemblage of men, and which, through it, becomes a compact mass and a unit, infusing, by means of a skilful and ingenious system of mechanical movements, an extraordinary degree of mobility into all the different parts of which the army is composed.

2. The several parts forming the elements of this whole must have dimensions, a form, and limits, which are the necessary consequences of the faculties of the man and the arms he uses.

3. Nothing is arbitrary in the organization of troops and the movements of armies. Everything, on the contrary, should depend upon rules, which themselves are derived from certain laws. Their application at the right time and in the proper manner forms the whole science of military operations.

4. An army is composed of its material and the men. There are natural and determined relations between these two elements, which, however, vary with the circumstances and the end proposed. These proportions do not depend upon caprice, but solely upon the nature of things.

5. The greater or lesser efficiency of these two elements powerfully influences the result, and the quantity of each mostly depends upon its quality.

6. A third element influences the value of troops—it is the moral

element. It alone often surpasses all others in importance, although the latter influence, to some extent, the efficiency of a body of troops, because it is requisite that the body be in existence before the spirit can animate it.

Thus, beyond a certain limit, the real strength of an army is not augmented by reason of the number of its soldiers and the resources in material, but much more by reason of the spirit by which it is animated.

7. To develop the spirit of an army, to augment its confidence, to speak to its imagination, to exalt the soul of the soldier—such should be the constant object of a general's cares and efforts.

8. Military spirit has for its element the *esprit de corps;* it is a powerful resort which can never be fostered too much. In the opinion of each soldier the army to which he belongs, and the general under whom he serves, should appear invincible; he should constantly maintain that his division is the best one in the army, and his regiment the bravest and most glorious one in the service. With these convictions, his strength and his courage will be tenfold augmented.

9. Lastly, every warrior should be profoundly imbued with the idea of his country's glory, and devotion to his chief magistrate, who is the representative and exponent of its grandeur. He should constantly remember that love for his country, the divine sentiment by Providence graven into the heart of every human being, will sustain him always, will make him a great man, and that it will place him above all eventualities. But this sentiment should not be a vain expression—it must be sincere, serious, and energetic; its reality should be proven, whenever necessary, by every sacrifice. History of every age has transmitted to us examples; even if they are rare, they are sublime, and their results have astonished mankind.

10. The best army is the one, therefore, which most fully satisfies the conditions above enumerated; their "*ensemble*" and their accord constitute its true value. As these conditions are almost ever changing and of difficult appreciation—because the mind can not embrace all possible combinations at once—no one can in advance determine the effective power of an army in any rigorous and definite manner; we can only judge of an army before it has been tried, by a sort of instinct which is not very far from the truth. But at a later period this value can, with certainty, be determined by the nature of the performances of an army, and their results.

My labors come here to a close. This sketch suffices for the fulfilment of the object I have had in view. To give to each part of which it is composed all the finish of which it would be susceptible, requires too extended labors, which I have neither the strength nor the wish to undertake. I have, however, said enough to lead the minds of soldiers to reflect upon their calling, and to show to them that our sublime profession is based upon certain principles which ought never to be disregarded; and that when they are respected, they impart the greatest possible value to the means of action they control, which, in fact, should be the constant aim of him who is called upon to command.

APPENDIX.

EXPOSITION OF THE GENERAL PRINCIPLES OF THE ART
OF WAR.

A new translation of General JOMINI's celebrated Thirty-fifth Chapter,
of Part First, of "*Traité des Grandes Opérations Militaires.*" From
the French, by Colonel FRANK SCHALLER, Confederate Army.

There have existed, in all times, fundamental principles upon which
the correct combinations of war repose, and to which they should all
be compared, to enable us to judge of their true merit.

These principles are immutable, independent of the kind of arms, of
the periods, and of the places. Genius and experience indicate the variations of which their application is susceptible. For thirty centuries
there have been generals who have, more or less happily, applied them.
Cyrus and Hannibal were great captains; Greece and Rome have furnished several; Alexander oftentimes manœuvred with skill; Cæsar
did not less successfully lead in wars of invasion and great operations;
Tamerlane evēn, of whom we know so little, has left institutions which
illustrate, upon every page, that natural genius which knows how to
command men and to triumph over every obstacle. When we compare the causes of the victories of antiquity with those of modern times,
we are quite surprised to find that the Battles of Wagram, of Pharsalia,
and Cannæ were gained through the same first cause.

However, through some fatality difficult to be comprehended, the
greater part of writers who have treated upon the military art appear
to have banded together in the endeavor to seek, in a thousand accessory details, the proper direction of great operations, or the wise employment of masses on the day of battle. The result has been a host of
works in which the authors by arranging, in their own manner, insignificant details, have undoubtedly proven a great deal of mind and
erudition, but at the expense of obscuring a science which they had
the intention to place within the reach of every one; several of them
have gone so far even as to devote entire chapters of works entitled
"*The Art of War*" to the manner in which officers should carry their
sword, and to the shape of gun-ramrods.

23

The result of these fatiguing dissertations has been to persuade many military men, otherwise very estimable, that there are no rules whatever in warfare; an absurd and unsustainable error.* Undoubtedly, there exists no *system* of war exclusively good, because they all are the results of hypothetical calculations. A system is a combination of the human mind, subject to deceive itself, and which oftentimes, by the aid of high-sounding phrases and technical words artfully arranged, gives to the falsest ideas the color of truth. But it is entirely different with *principles*. They are invariable; the human mind can neither modify nor destroy them.

To give exact notions of war it was, therefore, necessary that authors, instead of creating absurd systems, which destroy each other, should have begun by establishing the principles with which all combinations may be reconciled. This would have been a greater and more difficult labor—but one which would have secured some fixed result. We would not now find so many persons incredulous about the real state of the science. Mack would not have written, in 1793, that the longest lines were the strongest; Bülow, in his chapter of eccentric retreats, would not have pretended that a beaten army, to save itself, ought to be divided into as many corps as there are roads to retreat upon, should it even never succeed in reassembling its columns thus broken; neither would a system of cordon have been introduced, which scatters an army upon all the different roads, at the risk of being taken, as Turenne did with that of Bournonville in Alsatia.

Frederic had wisely written that the talent of a great captain consisted in forcing his enemy to divide; and, fifty years thereafter, several generals thought it an admirable manœuvre to divide their own forces as much as they could. Such a subversion of ideas could not fail to be the consequence of that uncertainty which controlled individual opinions; in fact, the grossest errors would not have thus been advanced, and the most eminent truths of the art would not have been misconstrued by military men, if, instead of laying down vague suppositions and uncertain calculations, military writers had endeavored to demonstrate incontestable principles, and 'to give a common regulator to opinions till then divergent.

---

* I heard a general of a certain reputation say, at the Château of Austerlitz, when speaking of a cavalry charge: "I would like very much the famous tacticians to explain to me by what rule we came out of that charge, where the squadron of both parties were mixed together." Undoubtedly, in a *mêlée* of cavalry, where the troops are already too much engaged to think of manœuvring, the only rule is to sabre; but is there anything proved by this truth? What was that charge by itself in the grand *ensemble* of the battle? Napoleon, who ordered it, has explained it already; it was the action of a secondary mass to restrain an effort of the enemy, while the great blow was struck upon a different portion of the field.

I have dared to undertake this difficult task without, perhaps, possessing the necessary talent for its complete solution: but it has appeared to me of importance to lay down the basis, whose development would have been much longer postponed, had we not been enabled to profit by circumstances in order to establish them.

The only means to arrive at my aim was first to indicate the principles, and then to show their application and their proofs by the history of twenty celebrated campaigns. This history [General Jomini is here speaking of his history of the wars of Frederic II, the wars of the Revolution, and of those of the Empire] should, then, present a strong and rational criticism of every operation which would be contrary to established rules. Had I approved of what was in opposition with these rules, I would have been guided by blamable and unworthy motives in the work I have undertaken. Whatever were the personal qualities of a general and the reputation which he enjoyed, I have frankly revealed every fault he has committed; I have not even hesitated an instant to come into collision with my private affections. After such an avowal, my reflections can neither be attributed to envy nor to personal enmity; the cause will be entirely in the interest of the art.

The fundamental principle of all military combinations consists *in making, with the largest mass of one's forces, a combined effort upon the decisive point.*

It will be well understood that a skilful general can, with sixty thousand men, beat one hundred thousand men, if he succeeds in throwing fifty thousand men into action upon a single portion of the line of the enemy. The numerical superiority of the troops not engaged becomes, in such a case, more hurtful than advantageous, because it can not but increase the disorder, as the Battle of Leuthen has proved.

The means of applying this maxim are not very numerous; I am going to attempt to indicate them :

I. The first means is to take the initiative of the movements. The general who succeeds in placing this advantage on his side is master, to employ his forces where he deems it suitable to carry them; the one, on the contrary, who awaits the enemy can not be master of any combinations, since he subordinates his movements to those of his adversary, and because he has it no longer in his power to arrest the latter's designs when they are in full execution. The general who takes the initiative knows what he has to do; he conceals his march; surprises and overwhelms an extremity and a weak part. The one who waits is beaten upon one of his parts, before he is even informed of the attack.

II. The second means is to direct one's movements upon that weak part which it is most advantageous to carry. The choice of this part depends upon the position of the enemy. The most important point will always be the one whose occupation will procure the most favorable chances and the greatest results. Such will be, for example, those positions which aim at gaining the communications of the enemy with the base of his operations, and to crowd him back upon some insurmountable obstacle, such as a sea, a large river without any bridge, or a great neutral power.

In double and disunited lines of operations, it is requisite to direct one's attacks upon the points of the centre; by throwing thither the mass of one's forces the isolated divisions which guard those points are overwhelmed. The scattered corps upon the right and the left can no longer operate in concert, and are forced to make those eccentric retreats, whose terrible effects have been felt by the armies of Wurmser, of Mack, and of the Duke of Brunswick. In simple lines of operations and in contiguous lines of battle the weak points are, on the contrary, upon the extremities of the line. Indeed, the centre is within reach of being simultaneously sustained by the right and the left; while one extremity, when attacked, would be overthrown before sufficient means could arrive from the other wing to support it, because its means would be much too far distant, and could not be employed except one after another.

A deep column, attacked upon its head, is in the same situation as a line attacked upon its extremity; they will both be successively engaged and beaten, as has been demonstrated by the defeats of Rossbach and Auerstedt. However, it is easier to make new dispositions in the case of a deep column, than it would be with a line of battle which would find itself attacked upon one extremity.

In executing, by strategy, a general movement upon the extremity of the line of operations of the enemy, not only can we act in masses upon a weak part, but we can, from this extremity, easily gain the rear and the communications with either the base or with the secondary lines. Thus Napoleon by gaining, in 1805, Donauwerth and the line of the Lech, had established his masses upon the communications of Mack with Vienna, which was the base of that general with Bohemia, and he made it impossible for him to join the Russian army, which was his most important secondary line. The same operation took place in 1806, upon the extreme left of the Prussians, about Saalfeld and Gera. It was again repeated in 1812, by the Russian army in its movements upon Kaluga and Krasnoï, and in 1813 by the Allies, when they directed their march across Bohemia, upon Dresden and Leipzig, against the right of Napoleon.*

---

* It has been remarked that central lines did not save Napoleon in the direction of Dresden in 1813, nor in the Champagne in 1814; but I may observe, in my turn, that he, nevertheless, owed to this system his momentary successes in both of these campaigns. The cause of his reverses consisted in the inequality of the strife and the secondary means; in the difference of the nature of his troops; in

III. The result of the preceding truths proves that, if it be requisite to attack by preference the extremity of a line, it is likewise necessary to beware of attacking both extremities at the same time, at least if one have not very superior forces. An army of sixty thousand men, which forms two corps of about thirty thousand combatants each, by attacking both extremities of an army equal in numbers, deprives itself of the means of striking a decisive blow by uselessly multiplying the number of the means of resistance which the enemy can oppose to its two detachments. It even exposes itself, by an extended and disunited movement, to the assemblage of its adversary's masses upon one point, and to be annihilated by the terrible effect of their superiority. Multiplied attacks upon a larger number of columns are still more dangerous, and more contrary to the great principle of the art, especially as they can not enter into action at the same instant and upon the same point.

It follows from this maxim that it is suitable, on the contrary, when we have masses much superior to those of the enemy, to attack upon both extremities; we thus succeed to bring into action more men than he upon each one of his wings, while by keeping very superior forces upon one single point the adversary might deploy his, and force us to combat with equal numbers. Care should be taken in such a case to throw the heavy portion of one's forces upon that wing where the attack would promise the most decisive results. This we have demonstrated by the relation of the Battle of Hochkirch, in the Seven Years' War.

---

placing Bohemia and Bavaria in the rear of his extreme right, and, so to say, upon his own communications. For the rest, I may still add that the system of central masses had not been applied until then but by armies of one hundred and fifty to two hundred thousand men at most, and as it would be useless to concentrate more forces upon the same line, since it is already a difficult matter to engage as many troops on the same day and upon the same battle-field.

Nor have I given any exclusive preference to central operations, since I have often presented those upon one extremity of the enemy's line as most advantageous. Besides, we must not confound a central line of operations opposed to two parties upon one single front (for example, that of Archduke Charles against Moreau and Jourdan in 1796), with a line of operations totally surrounded by enemies; these last are much less favorable; they may even became dangerous, when the masses of the enemy are more numerous.

Finally, I may say, in resuming, that one mass surrounded by all Europe risen against it, composed of heterogeneous parts, famished by its own greatness, and by light troops such as have never before been seen, could not have, by means of a central position alone, escaped the fate which struck Napoleon in Saxony. But one exception alone does not destroy a rule or a general maxim: and, in all ordidary wars, a power which will fight with equal chances—that is to say, with equal means—by applying this system will inevitably triumph, if its enemies follow a contrary system. I appeal to the most distinguished general officers of all armies, and cite as proofs the best feats of arms in modern history.

IV. To make a combined effort with a large mass upon one single
point, it is requisite, in strategic movements, to keep one's forces upon
a space very nearly square, that they may be more disposable.*   Large
fronts are as contrary to good principles as broken, disunited lines,
great detachments, and isolated divisions, unable to maintain themselves.

V. One of the most efficient means to apply the general principle
which we have indicated is, to make the enemy commit faults contrary to that principle.   With several small bodies of light troops, we
can give him uneasiness upon several important points of his communications.   It is likely that, not knowing their strength, he will oppose
to them numerous divisions, and break up his masses; these light troops
contribute likewise to protect the army perfectly against any kind of
danger.

VI. It is very important, if we take the initiative of a decisive movement, to neglect nothing concerning the enemy's positions, and the
movements which he would be enabled to make.   Espionage is a useful
means, whose perfection can not be too highly estimated; but what is
still more essential is, to have always a perfect system of reconnoissances by partisan rangers.   A general should scatter small parties in
every direction, and their number must be increased with as much care
as this system must be avoided in great operations.   To this end several divisions of light cavalry will be organized, which never enter into
the lists of combatants.   To operate without these precautions would
be to march in darkness, and to be exposed to disastrous chances which
a secret movement of the enemy would produce.   They have been too
much neglected, because the department of espionage has not been sufficiently organized beforehand, and officers of light troops have not always the necessary experience to conduct their detachments.†

---

* By this it must not be understood that it is necessary to form a regular squared
column, but that the battalions should be disposed upon the ground so as to be
able to march, with the same degree of promptitude, from all points toward the
one which would be assailed.

† The immense advantages which the Cossacks have given to the Russian
armies are a proof of the truth of this article, written in 1806.   These light troops,
insignificant in the shock of a great battle, are terrible in the pursuit.   They are
the most formidable enemy of all the combinations of a general, since he is never
sure whether his orders have been received and executed, as his convoys are
always compromised, and his operations uncertain.   As long as an army possesses
only a few regiments of them, their whole value is unknown; but if their number
is from fifteen to twenty thousand, their whole importance has been felt, especially
in countries whose population is not opposed to them.

As soon as they have once carried off a convoy, it is necessary to escort them all,
and that the escort be a numerous and well-conducted one.   We are never certain

VII. It is not sufficient, for the proper operations in war, to skilfully throw our masses upon the most important points ; we must also know how to engage them there. When we are once established upon these points, and remain there in inaction, then is the principle forgotten. The enemy is enabled to make counter-manœuvres, and, in order to deprive him of this means, it is necessary, as soon as we have gained his communications, or one of his extremities, that we march upon him and engage him. It is at that time particularly that a well-combined and simultaneous employment of our forces is of importance. The masses present do not decide any battles, but the masses in action. The first decide in the preparatory movements of strategy ; the last determine the success of the action.

To obtain this result, a skilful general must seize the instant when it becomes necessary to carry the decisive position of the battle-field, and he must combine the attack so as to engage all of his forces at the same time, with the sole exception of the troops retained to form the corps of reserves.

When an effort based upon such principles will not be successful in procuring the victory, it need not be expected from any other combination, and the only remaining hope will be to strike a last blow with this corps of reserves, in concert with the troops already engaged.

VIII. All the combinations of a battle may be reduced to three systems :

The first, which is purely defensive, consists in awaiting the enemy in a strong position, without any other aim than to maintain ourselves there ; of such a nature were the dispositions of Daun at Torgau, and of Marsin within the lines of Turin. These two events sufficiently demonstrate how vicious such combinations are.

The second system, on the contrary, is entirely offensive ; it consists in attacking the enemy wherever we can find him, as Frederic did at Leuthen and at Torgau, Napoleon at Jena and Ratisbon, and the Allies at Leipzig.

---

of any tranquil march, because we never know where the enemy is. These mean jobs require immense forces, and the regular cavalry is soon put "*hors de service*" by fatigues which it can not bear. The Turkish militia was nearly as effective against the Russians as the Cossacks were against other European armies ; the convoys are not more secure in Bulgaria than they were in Spain and Poland. As for the rest, I believe that in the other armies several thousands of volunteer hussars or lancers, raised in the beginning of the war, well-conducted, and operating where their chiefs would conduct them, would nearly answer the same end ; but they must always be considered as "*enfans perdus*," because, should they receive their orders from the adjutant-general's office, they would no longer be partisans. It is true they would not have the same qualities, and would not, in the long run, fight as well as Cossacks, but to an inevitable evil we must oppose every possible remedy.

The third system, lastly, is somewhat the mean between the two others; it consists in selecting a field of battle whose strategical conveniences and advantages of "terrain" are perfectly known, in order to await the enemy there, and to choose, during the day, the very moment most suitable for taking the initiative, and to fall upon the adversary with every chance of success. The combinations of Napoleon at Rivoli and at Austerlitz, those of Wellington at Mont-Saint-Jean and in most of his defensive battles in Spain, must be ranked in this class.

It would be a difficult matter to give fixed rules to determine the employment of these two last systems, which are the only ones suitable. Regard must be had to the moral state of the troops of each party, to the national character, whether it be more or less phlegmatic or impetuous, and, lastly, to the obstacles which are presented by the field of battle. It is therefore seen that these circumstances, above all others, must direct the genius of a general, and these truths may be reduced to the three following points:

1. That with troops well accustomed to war and upon ordinary ground, the absolute offensive or the initiative of the attack is always most suitable.

2. That upon grounds of difficult access, either by nature or owing to other causes, and with disciplined and submissive troops, it is perhaps more advantageous to permit the enemy to arrive in a position previously well known, in order to take thereafter the initiative against him, when his troops are already exhausted by their first efforts.*

3. That the strategical situation of the two parties may, nevertheless, require that we sometimes attack with strong force the positions of our adversary, without permitting any local consideration to prevent us from so doing—such are, for example, the circumstances in which it would be requisite to prevent the junction of two hostile armies, to fall upon a detached part of the enemy's army, or upon an isolated corps beyond a stream, etc.

IX. The orders of battle, or the most suitable dispositions to conduct troops into combat, should have for their object to give to the troops at the same time mobility and solidity. It appears to me that, to satisfy both of these conditions, troops remaining upon the defensive might be partly deployed and partly in column, as was the Russian army at Eylau; but the corps disposed in order to attack a decisive point must be composed of two lines of battalions; each battalion, instead of being deployed, would be formed in column by divisions in the following manner :†

---

* The Battle of Kunersdorf, which offers many points of resemblance with that of Mont-Saint-Jean, additionally justifies this train of reasoning.

† A division of two platoons: thus, the battalion having six companies, or six platoons, will have three divisions; which, in fact, will form it upon three lines.

6th.	5th.	4th.	3d.	2d.	1st batt.
———	———	———	———	———	———
———	———	———	———	———	———
———	———	———	———	———	———

12th.	11th.	10th.	9th.	8th.	7th.
———	———	———	———	———	———
———	———	———	———	———	———
———	———	———	———	———	———

This order offers infinitely more solidity than a deployed line, whose wavering prevents the impulsion so necessary for such an attack, and which hinders the officers from leading their commands. However, to facilitate the march, to avoid the too great depth of the mass, and to augment, on the contrary, the front, without at the same time lessening its consistence, I believe it to be suitable to place the infantry in two ranks. The battalions will thus become more mobile, since the march of the second rank, pressed between the first and the third, is always fatiguing, wavering, and in consequence less lively. They will, besides, have all the strength desirable, as the three ployed divisions will present six ranks in depth, which is already more than sufficient. Finally, the front, augmented by one-third, will have a greater fire, in case it should avail itself of it, and at the same time it will greatly impose upon the enemy, and, by showing to him more men, it will give less play to his artillery.

X. Upon grounds of difficult access, such as vineyards, enclosures, gardens, and entrenched heights, the defensive order of battle should be composed of battalions deployed in two ranks, and covered by numerous platoons of skirmishers. But the attacking troops, as well as the reserve, could not be better placed than in columns of attack in the centre, as we have indicated it in the preceding article, because the reserve, before being ready to fall upon the enemy at the decisive moment, should be prepared to do it with force and vivacity—that is to say, in columns.* This reserve may, however, be partly deployed until

---

* It has been said that Lord Wellington fought nearly always deployed. This may be true in the case of troops which remain always upon the defensive; but in the case of offensive and manœuvring wings, I believe that it is necessary to form columns. In the contrary case, it will be entirely the fault of the beaten army if it permits itself to be conquered by such a system, because a general ought to desire no better chance than to have an adversary who always gives battle with deployed lines.

I once more appeal, in regard to this subject, to the generals who have been engaged in the great European wars. It remains only to add that, by considering one order of battle as the most advantageous one, it is not intended to say that

24

the moment of attack, in order to be imposing to the enemy by its extent.

XI. If the art of war consists in concerting a superior effort of one mass against the weakest parts, it is incontestably necessary to push a beaten army in a lively manner.

The strength of an army consists in its organization, and in the *"ensemble"* resulting from the bond of all parts with the central point which causes them to move. After a defeat, this *"ensemble"* exists no longer; the harmony between the head which combines and the corps which should execute is destroyed—their relations are suspended and nearly always broken. The entire army is one weak part; to attack the same is equivalent to marching to a certain triumph. What proofs of these truths we find in the march upon Roveredo and the gorges of the Brenta, to complete the ruin of Wurmser; in the march from Ulm upon Vienna—in that from Jena upon Wittenberg, Custrin, and Stettin!* This maxim is often neglected by mediocre generals. It seems that the whole effort of their genius and the goal of their ambition is limited to become masters of the field of battle. A victory of this kind is nothing but a mere shifting of troops without any real utility.

XII. To render decisive the superior shock of one mass of troops, it is necessary that the general do not in any degree attend less to the moral condition of his army. What would, in fact, be the value of fifty thousand men placed in line of battle before twenty thousand, if they were not possessed of the necessary impulsion to engage and overthrow the enemy? Not only the soldier is thereby affected—it influences still more particularly those who are called upon to lead him. All troops are brave when the chiefs give the example of a noble emulation and of a proper devotion. A soldier must not face the fire by reason of fear alone, which a rigorous discipline inspires; it is necessary that he cheerfully confront it, imbued with that self-love which impels him not to cede anything to his officers in point of honor and bravery, and especially through that confidence in the wisdom of his

---

victory would be entirely impossible, if it be not strictly applied; localities, general causes, superiority of numbers, the moral condition of the troops, and the generals, all are considerations to be weighed in the general aspect. And when we reason upon a general maxim, it must be admitted that every one of these chances are of equal weight.

* This chapter was published in 1806. Since then the Russian army has furnished a renewed proof of this truth, by the activity and the perseverance with which it followed up its successes toward the close of the year 1812. The Emperor Alexander has likewise made a brilliant application of this principle in 1814.

chiefs and the courage of his companions-in-arms with which his superior has succeeded in inspiring him.*

A general must be enabled to rely, in his calculations, upon the devotion of his lieutenants to the honor of the national arms. It is necessary that he be perfectly secure that a vigorous shock will take place wherever he has ordered that it be executed. The first means to arrive at this end is to make himself beloved, esteemed, and feared; the second means is to entrust into the hands of this general the choice and the fate of his lieutenants. If the latter have arrived at their respective grades by the right alone which seniority confers, it may be decided beforehand· that they scarcely ever will be possessed of the necessary qualities to completely discharge the important functions of their rank. This circumstance, by itself, may lead to the failure of the very best conceived enterprises.

It is seen, by this rapid exposition, that the science of war is composed of three general combinations, each of which offers but a small number of subdivisions and chances of execution. The only perfect operations would be those which would present the application of these three combinations, since that would be the permanent application of the general principle before indicated.†

---

*The rules undoubtedly vary according to the different nations, and all the lesser lights and shadows illustrative of the point of honor are not applicable to every army, as the Austrian military journal has with reason remarked, when speaking of one of my chapters. But whatever this journal may advance in this respect, it is certain that the rigors of discipline did not alone make the legions of Suwarrow so brave, but because he had the talent to electrify them in his own manner. Despite the criticism of the writer of that article, I still persist in believing that corporeal punishment is entirely unfit to act as an incentive. Its effects may be to modify, soften, and amend the disposition of the soldier, but it will never make him a good one, any more than the declamations which have, perhaps, become too general against this punishment. There are other means to excite the *morale* of an army, and I will cite an example for illustration. At the affair of Culm, a sergeant of the regiment of Devaux, when delivering to Prince Schwarzenberg a standard which he had taken, explained to this marshal the re-entering and salient angles, formed by a streamlet and the village which had been attacked by the corps of Colloredo. An officer of engineers could not have spoken better, and the prince himself was struck by it. This brave man had been a non-commissioned officer for nine years; he was handed two ducats (about six dollars), and the hope of a medal was held out to him for his standard, and his truly didactic narrative. Ought he not have merited another recompense, and was he not fit to command a company?

† National wars, where we are obliged to combat and conquer an entire people, alone make an exception to these rules; in wars of this kind it is a difficult matter to bring a people to subjection without dividing our forces—since, whenever we assemble them to give battle, we expose ourselves to lose the provinces already conquered.

The means to guard against these inconveniences is to have an army continually

The first of these operations is the art *of encompassing the lines of operations in the most advantageous manner;* this is precisely what is commonly and improperly called *a plan of a campaign.* I really do not see what is meant by this denomination, since it is a matter of impossibility to make a general plan for the whole of a campaign, the first movement of which may overthrow the whole scaffolding, and in which it would be impossible to see beyond the second movement.

The second branch is *the art to carry one's masses, in the most rapid manner possible, upon the decisive point of the primitive line of operation, or upon the accidental line.* This is what is vulgarly meant by "strategy." But strategy is merely the means of execution of this second combination; its principles may be found in the above-mentioned chapters.

The third branch is *the art to combine the simultaneous employment of one's greatest mass upon the most important point of a field of battle;* this is, properly, the art of combats, which several authors have designated by the name of "order of battle," and which others have presented under the name of "tactics."

And here we have the science of war in a few words. It was for having forgotten this small number of principles that the Austrian generals have been beaten from 1793 until 1800 and 1805; it was for the same cause that the French generals lost Belgium in 1793, Germany in 1796, Italy and Swabia in 1799.

I need not observe to my readers that I have treated here of the principles only which relate to the employment of troops, or of the purely military part of the art of war; other no less important combinations are indispensable for the proper conduct of a great war, but they belong to the science of governing empires rather than to that of commanding armies.

To be successful in great enterprises it is not only requisite to calculate the state of the respective armies, but also that of the means of the second line, which is designed to act as a reserve, and to replace the losses of all kinds, both in *personnel* and *materiel.* It is also necessary to know how to judge of the interior state of the nations according to what they have already previously sustained, and of the relative situation of their neighbors. Nor is it any less required to place into the balance the passions of the peoples against whom the war will be waged, their institutions, and the attachment they have for them. The situation of the provinces must likewise be taken into account, as well as the

---

in the field, and independent divisions for the purpose of organization in the rear. These divisions should be commanded by well-informed generals, good administrators, firm and just at the same time, because their individual labors may as much contribute to the submission of the provinces entrusted to their care as the force of arms.

distance of the power which it is intended to attack, because the disadvantages of the aggressor are multiplied in measure as he increases the depth of his line of operations. Lastly, it is necessary to judge of the nature of the country into which the war is going to be carried,* and the solidity of the alliances which may be made for a distant enterprise.

In a word, it is indispensable to know that science, a mixture of politics, administration, and of war, the bases of which Montesquieu has so well laid down in his work upon the causes of the grandeur and the decadence of the Romans. It would be difficult to assign to this science any fixed rules, and even any general principles; history is the only school in which we may find some good precepts; and it is likewise very rarely that we encounter any circumstances which resemble each other sufficiently to be taken for our guidance at a certain epoch, and to shape our actions according to what was done several centuries before. The passions of men exercise so great an influence upon events as to make it a matter of impossibility to prevent the failure of any undertaking, even when others, in similar circumstances, have succeeded.

Napoleon may, perhaps, have known this science, but his contempt for men caused him to neglect its application. It was not any ignorance of the fate of Cambyses or of the legions of Varus which caused his reverses; nor was it any forgetfulness of the defeat of Crassus, or of the disaster of Emperor Julian, and the result of the Crusades; it was the opinion which he had that his genius would secure to him incalculable means of superiority, and that his enemies, on the contrary, possessed none whatever. He is fallen from the pinnacle of his grandeur for having forgotten that mind and strength of man have also their limits, and that, the more enormous are the masses set in motion, the more is the power of genius subjected to the imprescriptible laws of nature, and the less it controls events. This truth, which has been demonstrated by the results of the affairs of Katzbach, Dennewitz, and even of Leipzig, would by itself make an interesting subject of study.

It does not enter into my plan here to repeat the important precepts which Montesquieu and Machiavel have left to us upon the great art of directing the movements of empires. In the course of the narrative of these celebrated campaigns some reflections will be suggested upon the change which the wars of the Revolution have caused in the ideas upon the organization and the manner of displaying national forces, upon their employment, and the consequences which will probably result therefrom in the future revolutions of the body-politic. Armies are no longer composed to-day of troops recruited voluntarily from the superfluity of a too numerous population; they are nations entire, call-

---

* This led me to write, as long ago as 1805, volume v, chapter iv, that the system of Napoleon was not applicable either to Russia or Sweden,

ed to arms by law; and they no longer fight for some demarcation of a frontier, but, in some sort, for their existence.

This state of things throws us back to the third and fourth centuries, and recalls to us those shocks of immense peoples which disputed among each other the European continent; and if legislation and a new international law do not interpose to limit these levies *en masse*, it is impossible to foresee where these ravages may stop. War will become a plague more terrible than ever, since the population of civilized nations will be gathered in, not, as in the Middle Ages, in order to resist savage hordes and devastators, but for the sorry maintenance of a political balance, and in order to determine, at the end of a century, whether such and such a province ought to have a prefect from Paris, Petersburg, or from Vienna, who would govern it with about the same laws and usages. It would be, however, high time that the cabinets should arrive at more generous ideas, and that henceforth blood should no longer flow save in the defence of the great interests of the world.

If this wish, truly European, must be consigned to banishment with the beautiful dreams of perpetual peace, let us deplore the little passions and interests which lead enlightened nations to cut each others' throats more pitilessly than barbarians; let us deplore the progress in arts, sciences, morals, and politics, which, far from bringing us any nearer to the perfection of the social state, appear to carry in their train the centuries of the Huns, the Vandals, and the Tartars.